THE FURTHER READING LIBRARY

LOÏE FULLER
LECTURE ON RADIUM

THOMAS WILFRED
CLAVILUX AND LUMIA HOME MODELS

RICHARD SHARPE SHAVER
SOME STONES ARE ANCIENT BOOKS

RICHARD FOREMAN
NO TITLE

MARGARET WATTS HUGHES
SOUND MAY BE SEEN

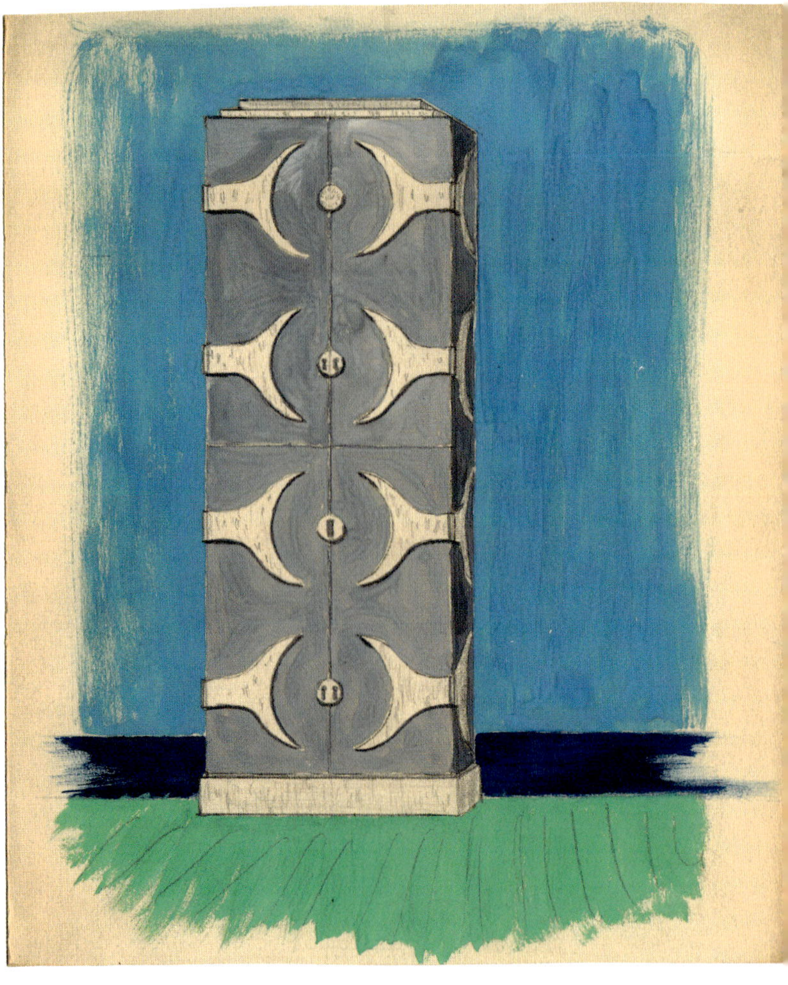

THOMAS WILFRED

CLAVILUX AND LUMIA HOME MODELS

THE FURTHER READING LIBRARY

CHRISTINE BURGIN BOOKS

SERIES EDITORS: CHRISTINE BURGIN AND ANDREW LAMPERT

Published by Christine Burgin Books
Ghent, New York

Special thanks to Skip Elsheimer, Carol and Eugene Epstein, AJ Epstein, Donna Stein, Keely Orgeman, Joshua White, Yale University Art Gallery, and the Manuscripts and Archives of Yale University Library.

ISBN: 978-0-9976456-3-7

Published in 2025

www.furtherreadinglibrary.com

Designed and typeset in Caslon by Laura Lindgren
Image preparation by Jason Burch
Front cover lettering by Craig Pettman
Printed on 115 gsm Kasadaka White
Printed in China

FRONTISPIECE: *Unrealized design for a* Clavilux Junior *with decorative hinges, n.d.*

CONTENTS

PUNCH and JUDY THEATRE

THURSDAY APRIL 18th at 3 P. M.

RECITAL IN COSTUME

BY

THOS. WILFRED

THE LUTE PLAYER

Old Mediaeval
and Elizabethan
Songs and Ballads

and

SOLO PLAYING ON
THE OLD 12 STRING
ARCH-LUTE.

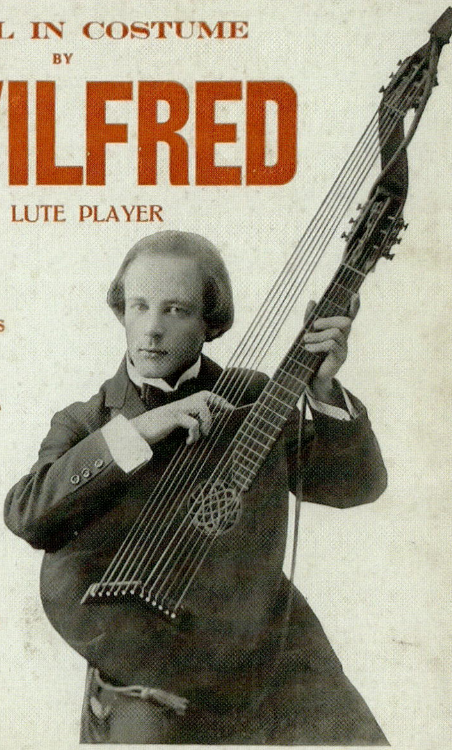

The entire tier of boxes
have already been sold.

ORCHESTRA SEATS

$2.00

at Box Office

155 West 49th Street

FORM, COLOR, AND MOTION

Doug Skinner

Thomas Wilfred devoted his life, with remarkable persistence, to the creation of a new art form. He called it *Lumia*: "The use of light as an independent art-medium through the silent visual treatment of form, color, and motion in dark space with the object of conveying an aesthetic experience to a spectator."[1]

Richard Edgar Løvstrøm (his original name) was born in Næstved, Denmark, in 1889. His mother died when he was two; his earliest memory was being taken into his father's office shortly after: "The sun was shining in there, and father to cheer me up took a crystal egg mounted on a letter press and moved it in the sunbeam. And the ceiling suddenly became covered with little traveling dots of spectral color, and father said I screamed with delight."[2]

As a child, Wilfred took art and piano lessons but never forgot those traveling dots, and he kept tinkering with cigar boxes, lightbulbs, and colored glass. After further art studies in Paris and London, however, he decided on a career in music.

So, in 1913, he changed his name to Thomas Wilfred and started touring as a singer, accompanying himself on the lute. Billed as "Wilfred the Lute Player," with long hair and

"Thomas Wilfred the Lute Player," poster, 1918

performing in a period costume, he claimed to be the last of the troubadours and one of only three lutenists in the world. Despite (or because of) this hyperbole, he toured extensively, performed for British and Danish royalty, and recorded three unreleased selections for Victor.[3] After his requisite stint in the Danish army, he moved to the United States. Throughout all of this, he kept working on his light machines.

Wilfred wasn't the first to propose an art of light. He cited Pythagoras and Aristotle, the fireworks of the Emperor K'ang Hsi, and the colored lights in Scriabin's tone poem *Prometheus*. Earlier instruments included Louis Bertrand Castel's Ocular Harpsichord (1725), Bainbridge Bishop's Color Organ (1877), and Alexander Rimington's Color Organ (1894). The flamboyant Lebanese pianist Mary Hallock-Greenewalt built an instrument she called the Sarabet (after her mother) and started performing on it in 1916, as well as writing and lecturing about her new art of color music, Nourathar.

There was also Claude Bragdon, an architect steeped in Theosophy and theories about the fourth dimension. He built several versions of his Luxorgan; more publicly, he designed the lighting for eight Festivals of Song and Light, massive outdoor singalongs in New York from 1915 to 1918, including one in Manhattan's Central Park. Wilfred was touring the area at the time; he must have heard about those thousands of people chorusing Stephen Foster in the parks and crossed paths with Bragdon.

In 1918, Wilfred and Bragdon, along with the painter Van Dearing Perrine, joined forces to build color instruments. They

"Hue Circle," ca. 1940–50. The circular arrangement of the colors corresponds to their placement on the rheostat control dial of a Wilfred lumia device.

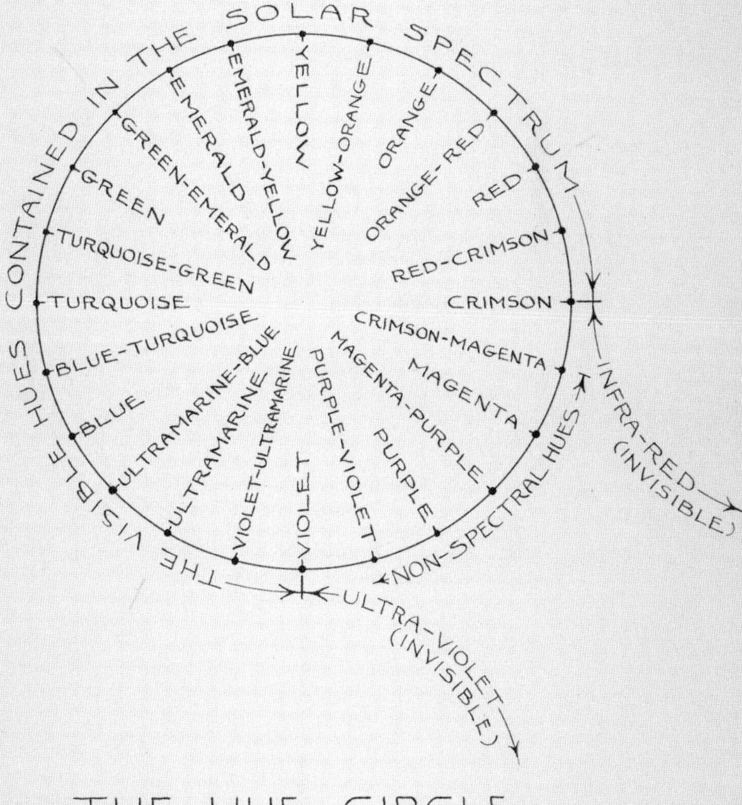

THE HUE CIRCLE

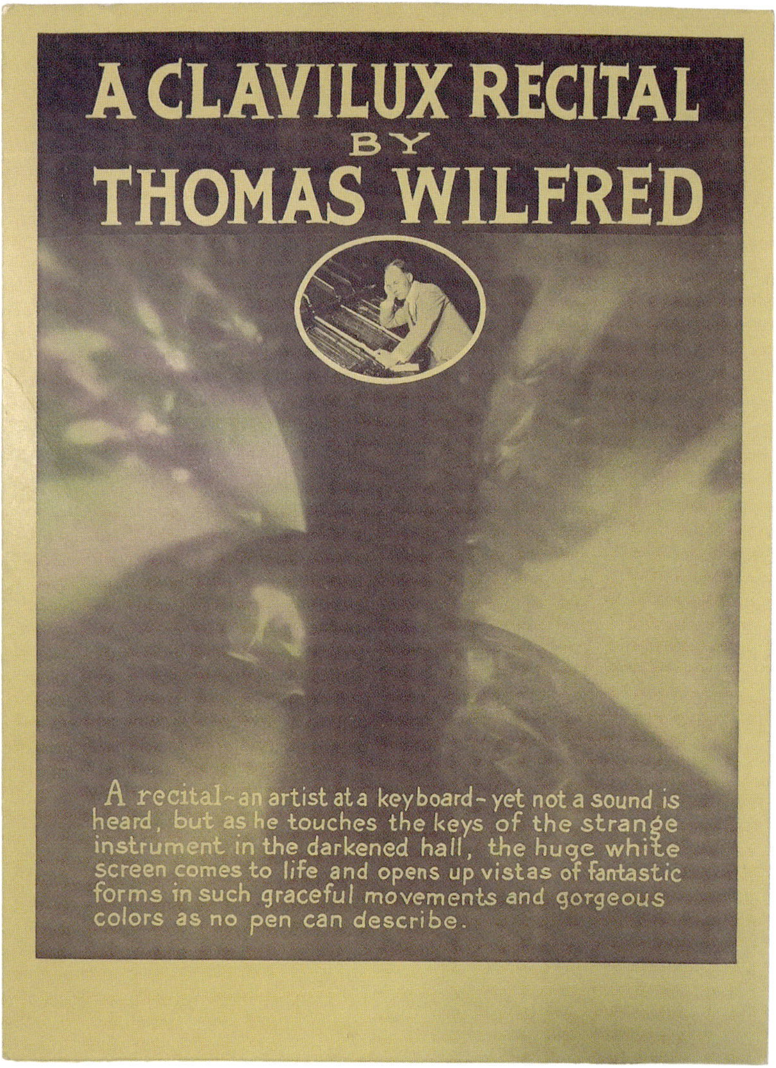
A CLAVILUX RECITAL
BY
THOMAS WILFRED
A recital - an artist at a keyboard - yet not a sound is heard, but as he touches the keys of the strange instrument in the darkened hall, the huge white screen comes to life and opens up vistas of fantastic forms in such graceful movements and gorgeous colors as no pen can describe.

called themselves, with a bow to Scriabin, the Prometheans. A patron of the Festivals, Walter Kirkpatrick Brice, invited Bragdon to build a color theater in his home in Huntington, New York.

Wilfred, the youngest of the three, promptly took over. As Bragdon later noted: "Although both Perrine and I had devised and exhibited mobile color machines, Wilfred, a skilled mechanic and electrician, with clear-cut ideas of what he wanted to do and how to go about it, soon assumed the dominant position in our triumvirate, with the result that Perrine and I were more or less relegated to the role of spectators rather than participants."[4]

Brice's support allowed Wilfred to concentrate on his machines. Previous instruments had shown fields of solid color, projected either by lights through tinted glass (as with Bishop's color organ) or by stage lights controlled by a dimmer (as with Hallock-Greenewalt's Sarabet); Bragdon had added decorative stencils to the screen or projector. Wilfred preferred a variety of shifting colors and forms. Earlier instruments were also meant to accompany music, usually matching pitches to colors—both Hallock-Greenewalt and Bragdon plotted out Beethoven's "Moonlight Sonata" note by note. Wilfred wanted visual music in its own right, played in silence. He denied any correspondence between colors and pitches and declared synesthesia a myth.

Wilfred built his first instrument, the Clavilux A (he hated the term "color organ"), in 1919. He created new effects by combining various "form, color, and motion-producing devices," like bulbs, lenses, reflectors, and painted glass plates.[5] Between

"A Clavilux Recital by Thomas Wilfred," poster, ca. 1926

1921 and 1937, he designed several portable versions of the Clavilux to take on tour.

In the next few years, Wilfred's life changed radically: he became an American citizen, married (to Ada Granberg, a Swedish actor), had two children, quit his singing career, and started touring with the Clavilux. He also fought off repeated legal challenges from Hallock-Greenewalt, who accused him of patent infringement for what she called his "kaleidoscope."

Just as Wilfred had once played a troubadour, he now posed as the pilot of a spaceship, telling his audience, "I hope to have transformed the hall into the cabin of a fantastic dream ship capable of traveling through space with the speed of thought."[6] The reviews were enthusiastic, but the audience limited: form, color, and motion are not always theatrical. Wilfred sought new applications for his work, but he rejected many proposals: "Advertising firms offered me tempting contracts for display and signboard use—Stockings, Chewing Gum, Laxatives, Cigarettes. To this day I shudder when I contemplate the harm I might have inflicted on lumia if I had yielded and sold it into slavery."[7] Despite his preference for silent lights, he did perform with musicians. One important appearance was a concert at Carnegie Hall in 1926, with Leopold Stokowski leading the Philadelphia Orchestra in Rimsky-Korsakov's *Scheherazade*. Even though Wilfred never matched colors to pitches, he did use thematic equivalents: "Each musical motif had a corresponding mobile form and color motif and these were blended and interwoven as the music was played."[8]

"Imagination—the Esthetic Concept; Reality—the Physical Equipment," ca. 1940–1950

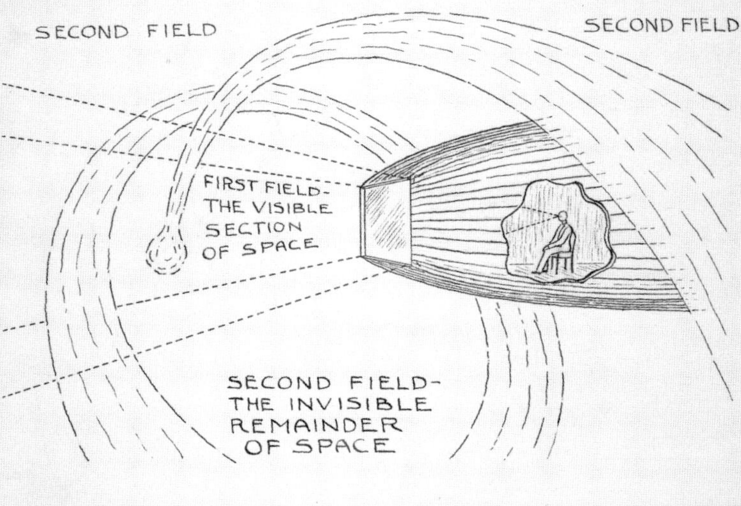

IMAGINATION — THE ESTHETIC CONCEPT.

THE ARTIST VISUALIZES A LUMIA COMPOSITION IN SPACE. HE AIMS TO PERFORM IT SO CONVINCINGLY SPATIAL THAT THE SPECTATOR IMAGINES HE IS SEEING IT THROUGH A LARGE WINDOW IN THE CABIN OF A MAGIC SPACE-LINER.

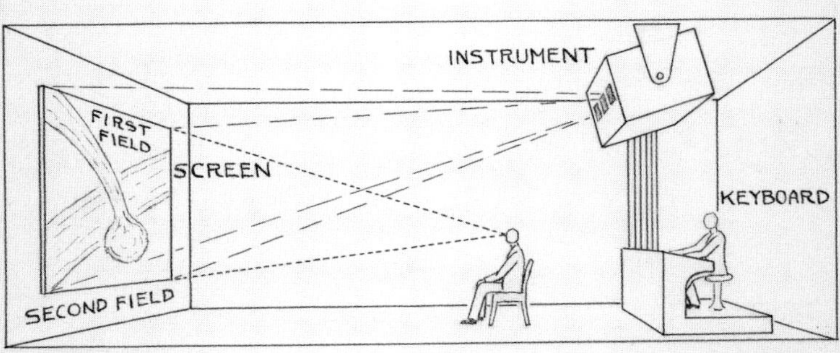

REALITY — THE PHYSICAL EQUIPMENT.

THE SPECTATOR IS SEATED IN A RECITAL HALL. THE MAGIC WINDOW IS A FLAT WHITE SCREEN. THE ILLUSION OF SPACE IS CREATED WITH A LUMIA PROJECTION INSTRUMENT CONTROLLED BY THE ARTIST FROM A KEYBOARD.

He also designed theatrical lighting and scenery (as had Bragdon) and created a number of light murals for hotels and office buildings.

Although he continued giving recitals, he had become dissatisfied with equating lumia with music, admitting that "attempts to design lumia instruments in imitation of musical ones will prove as futile as attempts to write lumia compositions by following the conventional rules laid down for music."[9] He started designing instruments as independent sculptures able to operate in homes or galleries on their own.

Some modifications were needed for the new instruments. Household wiring, for example, required lower wattage bulbs. There was also the problem of how to show the images. A screen was impractical, since he couldn't control its placement or the darkness of the room.

His first series, called Luminars, solved those problems by projecting images onto the ceiling through what appeared to be an ordinary lamp. A typical instrument, *Unit # 50, Elliptical Prelude and Chalice*, used two painted glass records that not only rotated but slid back and forth, as the light bounced off reflective cones. The result resembled a sort of small personal cyclone.

Later home instruments projected their images from behind onto a screen fitted in the front. Called Clavilux Junior, and later Home Lumia Instruments, they were housed in cabinets like radios or phonographs, looking like forerunners of televisions. Some were meant to be more interactive and included control panels, changeable color records, and blank records to be painted by the owner. Wilfred entered negotiations for wider distribution, including a possible unit combining a Clavilux with a radio and phonograph, but he kept his creations personal; he made only forty-two home instruments from 1928 to 1936.

In 1930, Wilfred founded the Art Institute of Light to preserve and promote his work. Three years later, he gave it a home in the Grand Central Palace in Manhattan. It wasn't the vast white cathedral he'd envisioned in drawings and models, but it was still spacious enough to build and store his instruments, teach students, and present regular recitals.

Many ideas for lumia, like a "silent visual carillon" set in a tower and "dry fountains" for parks, remained tantalizing drawings. The Fantascope was a notable exception. In the summer of 1939, Wilfred designed this therapeutic device for the Payne Whitney Psychiatric Clinic in New York. It was a sort of mobile Rorschach test, which played a forty-minute light sequence while recording patients' reactions. When the war forced him to close the Art Institute, he built a new studio in West Nyack, New York, while working for the Office of War Information.

Although he'd performed in galleries, his work wasn't displayed in them until the 1950s. Abstract Expressionism had become popular, and Wilfred's use of form, color, and motion fit with the work of younger painters. His first major exhibition was in *15 Americans*, at the Museum of Modern Art in 1952, which bracketed him with youngsters such as Jackson Pollock and Mark Rothko. Shows followed at the Stedelijk Museum in Amsterdam, the Walker Art Center in Minneapolis, and many others.

He returned to MoMA in 1963, when it commissioned his most complex and influential work: *Lumia Suite, Op. 158*, an

PAGES 16–17: *Drawing for patent 1,758,589, "Light Projection Display," filed by Thomas Wilfred on December 21, 1927, and issued May 13, 1930*

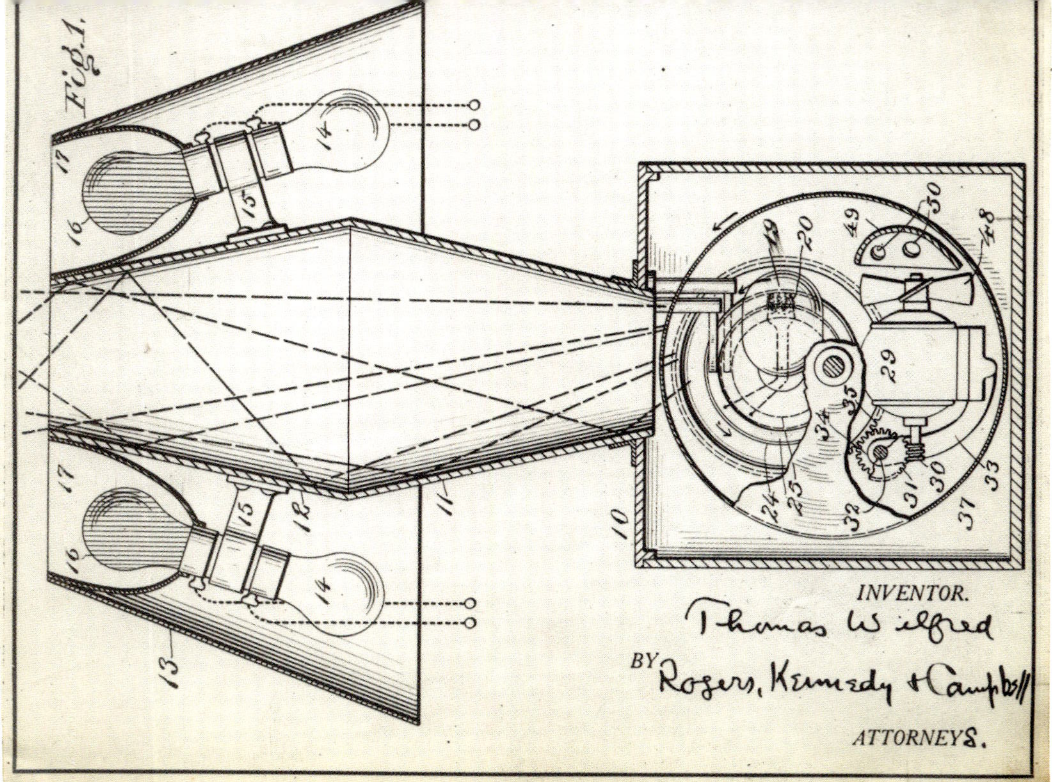
Fig. 1.
INVENTOR.
Thomas Wilfred
BY Rogers, Kennedy & Campbell
ATTORNEYS.

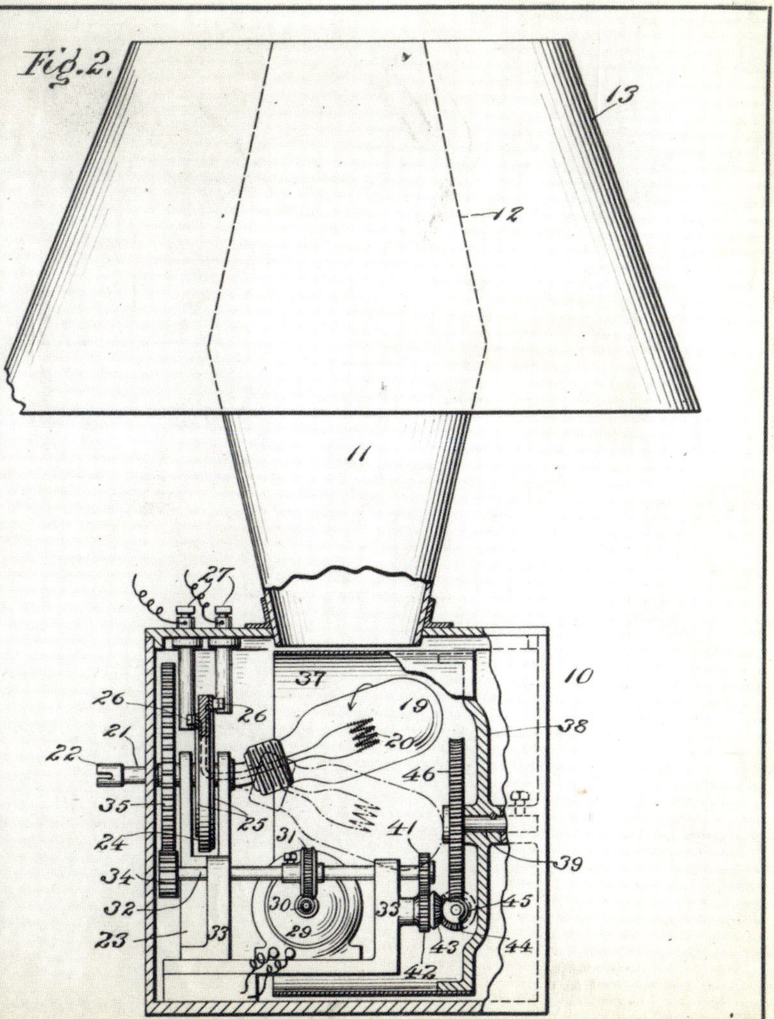
Fig. 2.
13
12
11
27
37
10
19
26
26
38
21
20
22
46
35
25
24
41
39
34
31
30
33
45
32
29
23
33
43
44
42

elaborate construction of projectors, reflectors, and filters, all whirring away behind a six-by-eight-foot screen and housed in its own little theater. It was his longest piece, designed not to repeat itself for 9 years, 127 days, and 18 hours. The museum's label was fittingly rhapsodic: "A descending arc slowly transforms the horizontal sequence into a central whorl of expanding and interlacing ellipses sweeping around a nucleus of intense changes."[10]

MoMA kept *Lumia Suite* running until 1980. By then the projectors were breaking down, the paint was peeling from the glass discs, and it had to be removed. It was seen by countless visitors, inspiring some to make their own versions of lumia—including light and space installation artist James Turrell and light show pioneer Joshua White, whose psychedelic Joshua Light Show let rock audiences listen to guitarists without having to look at them.

Wilfred's last years were devoted mostly to stage lighting. He designed lighting for theaters in Seattle, Georgia, and Fresno, marketed his Direct Beam and Multiplate projectors, and wrote *Projected Scenery: A Technical Manual.* He never abandoned lumia; his last work, *Luccata, Op. 162*, was completed in 1968.

In his last interview, in July 1968, he emphasized the expressive possibilities of lumia, noting, "The art form is peculiarly adapted to the unearthly, to the visionary," adding, "If it's just a display of changing colors, it means nothing to anybody." It was, he said, still in its infancy: "The Johann Sebastian Bach of lumia has not yet been born, or at least he may be in grade school now, and when the first real genius expresses himself through this art form with the instruments that are at his fingertips that he can express things through, then we may see something entirely different. But the art form is here to stay. We might as well get acquainted with it, because it is one of the great arts."[11] Wilfred died a month later, on August 15, at the age of seventy-nine.

NOTES

1. Thomas Wilfred, "Light and the Artist," *The Journal of Aesthetics and Art Criticism* 5, no. 4 (June 1947): 247–55.
2. Interview with Patricia Marx on radio station WNYC, July 18, 1968.
3. Victor recorded three pieces on November 11, 1918: "Sumer is icumen in," "Danish gypsy song," and "Bourrée."
4. Claude Bragdon, "My Experiments with Color-Music," *The Theosophist*, May 1936. Reprinted in *The Theosophist*, March 2013, 27–36.
5. Thomas Wilfred, "Clavilux," *Encyclopedia Britannica*, 14th edition, 1929, vol. 5, p. 784.
6. Thomas Wilfred Papers (MS 1375), Yale University Library.
7. Thomas Wilfred, "Light and the Artist."
8. Thomas Wilfred, "Stage Design: Visual Accompaniment in Light," *Encyclopedia Britannica*, 14th edition, 1929, vol. 21, p. 290.
9. Thomas Wilfred, "Composing in the Art of Lumia," *The Journal of Aesthetics and Art Criticism* 7, no. 2 (December 1948): 79–93.
10. Keely Orgeman, *Lumia: Thomas Wilfred and the Art of Light*, Yale University Art Gallery, 2017, p. 138.
11. Interview with Patricia Marx on radio station WNYC, July 18, 1968.

Doug Skinner has contributed to *Fortean Times*, *Strange Attractor Journal*, *Nickelodeon*, *Cabinet*, *Fate*, *Weirdo*, and *Typo*; his books include *Music from Elsewhere*, *The Potato Farm*, *Nominata*, and translations of Alphonse Allais, Charles Cros, Alfred Jarry, and Luigi Russolo.

FIRST TABLE MODEL CLAVILUX

The Luminar, 1928

THE HOME LUMINAR

Designed by the artist-craftsman Thomas Wilfred, creator of the now famous Clavilux. The home Luminar plays -automatically- the special color records inserted, like phonograph discs, into its mechanism.

Placed in the center of any room its powerful and pure mobile color images seem to open the ceiling to disclose the splendours of Arabian Nights in space beyond.

Gently drifting clouds change the room from minute to minute and the absence of all representation and realism permits the mind to travel as far as imagination can reach.

To a world of steam heat and indifferent electric light this new creation brings beauty and poetry far beyond that of the fireplace, which in past centuries inspired so many great works of art.

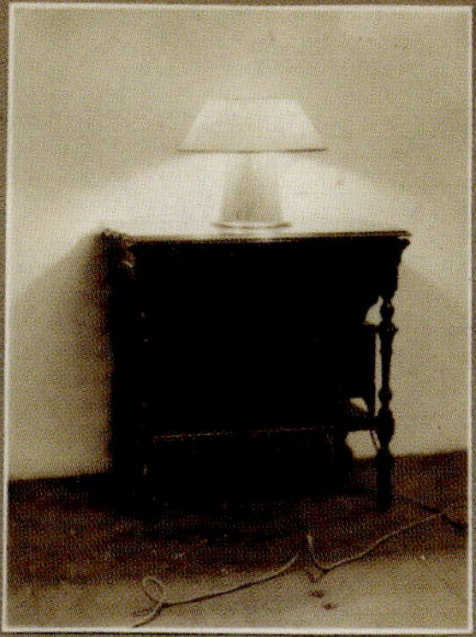

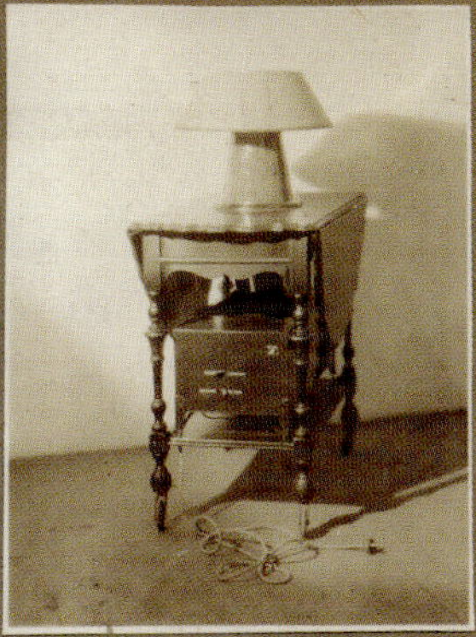

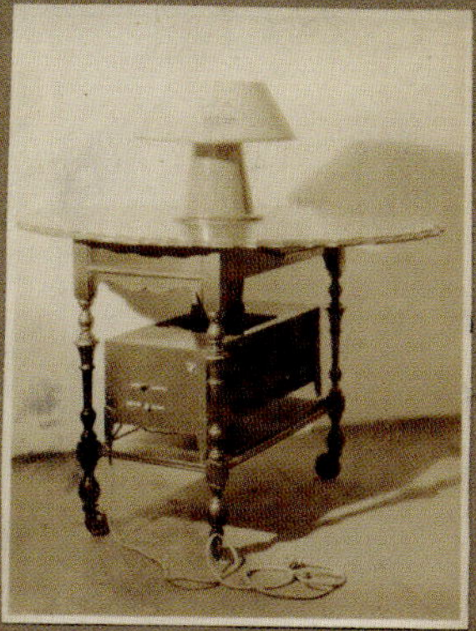

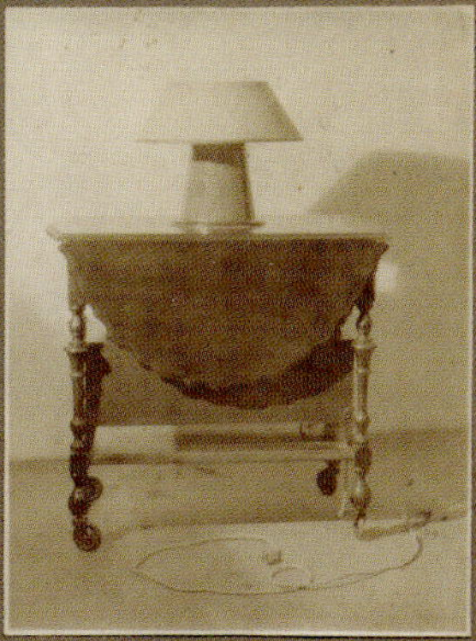

#36 "THE LUMINAR" 1928
AN AUTOMATIC CLAVILUX FOR THE HOME, FOR PRO-
JECTION ON A WHITE CEILING. TO BARKER DEC. 18, 1928
BACK DEC. 13. 1936 (11). LET DR. ADELSON FOR OCT. 1963 (100)

Thomas Wilfred spent most of the 1920s touring and performing in theaters throughout the United States and Europe with his portable Clavilux light organ. He designed The Luminar *to bring these acclaimed performances into the home. Wilfred housed an elaborate arrangement of mechanical gears, light-bulbs, painted records, and reflective surfaces inside of what appeared to be an ordinary table and lamp. At the flick of a switch one's living room ceiling could be transformed into a wondrous landscape of projected colored light.*

OPPOSITE: Unit #50, Elliptical Prelude and Chalice, *from the* First Table Model Clavilux (Luminar) *series, 1928. Collection Yale University Art Gallery, New Haven, Conn., Gift of Thomas C. Wilfred*

PAGES 24–25: *Images of two of several light projection possibilities created by* Unit #50

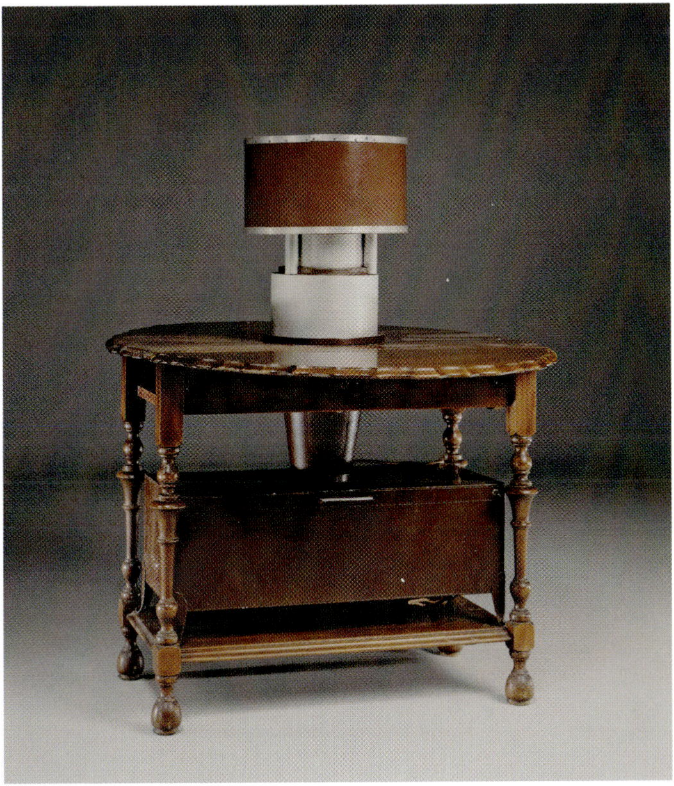

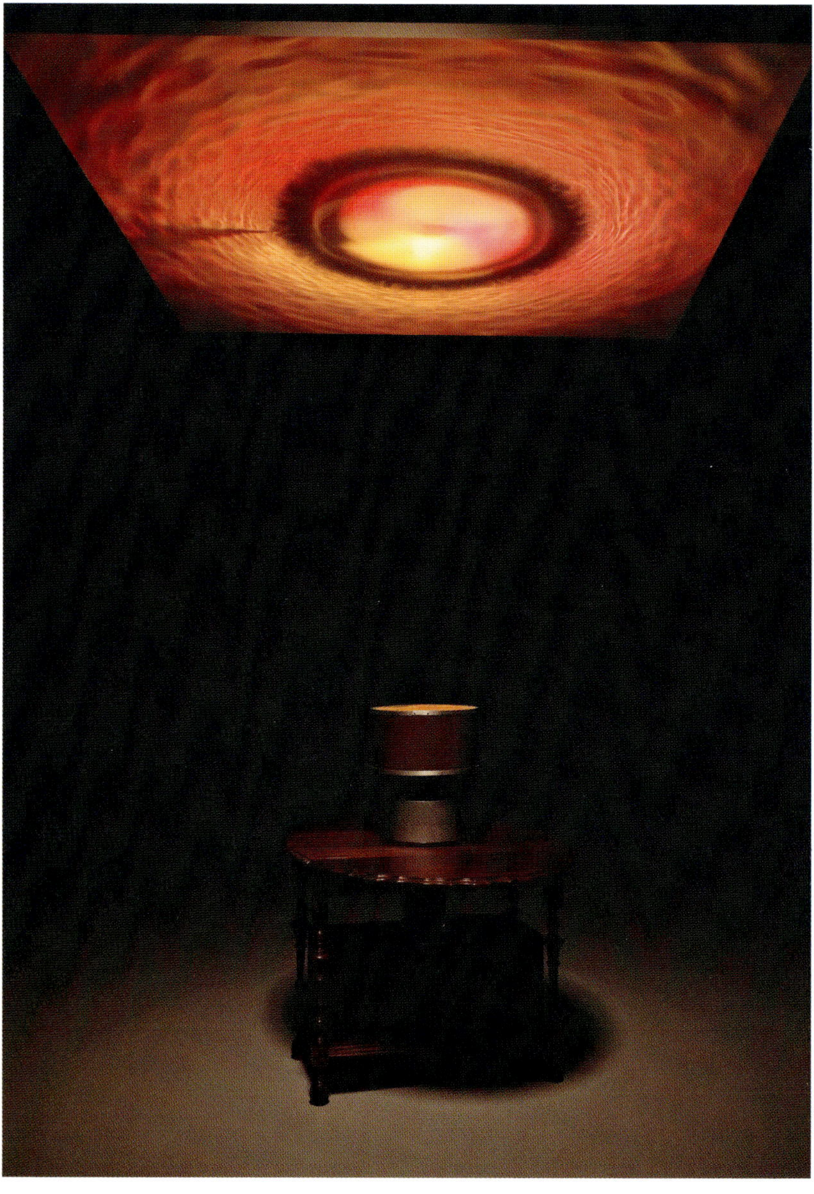

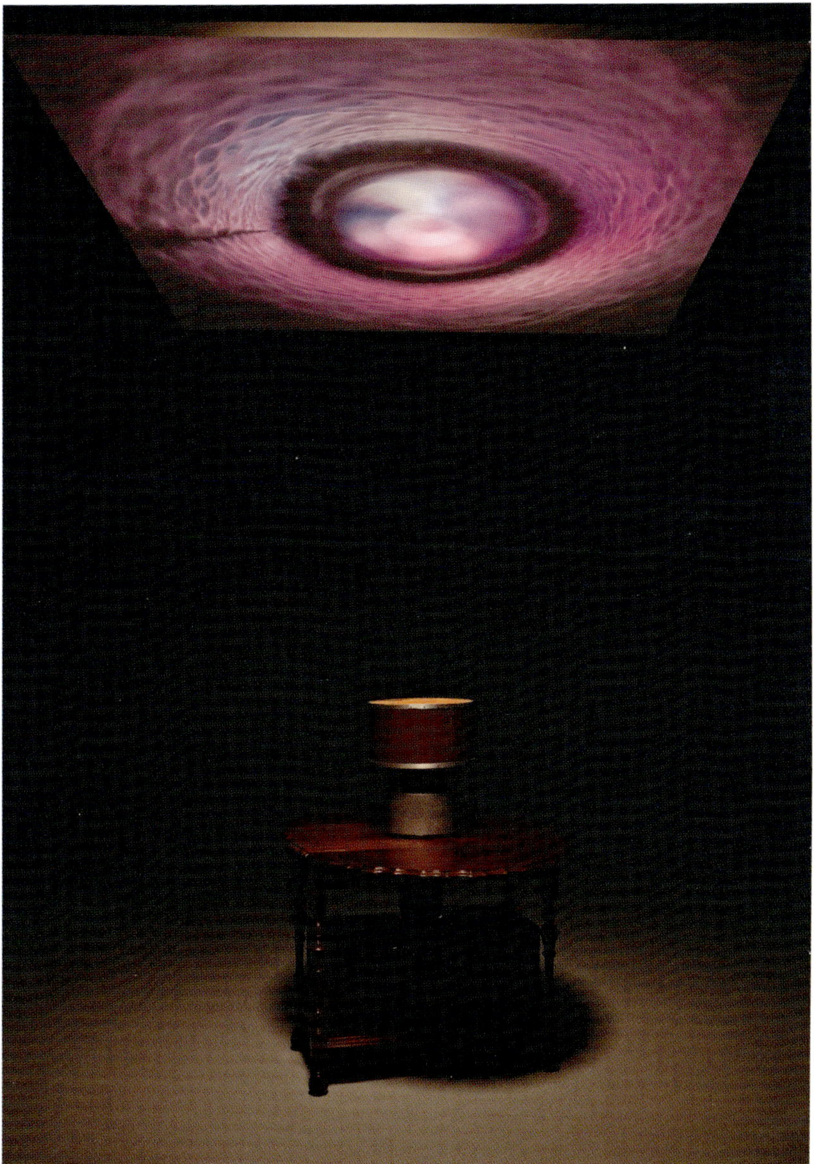

OPPOSITE: *Documentation of interior mechanism of* First Table Model Clavilux #36, *1928. Components from this unit were later repurposed by Wilfred for use in* Op. 83 Study in Depth *when shown in the 1952 MoMA exhibition* 15 Americans *(see pages 56–57).*

PAGES 28-31 AND 33-35: *Sketches for home Clavilux housing*

PAGE 32: *Documentation of both* Table Clavilux "Luminar" Model *of 1928 and the first home Lumia instrument to have a translucent screen built in 1935–36*

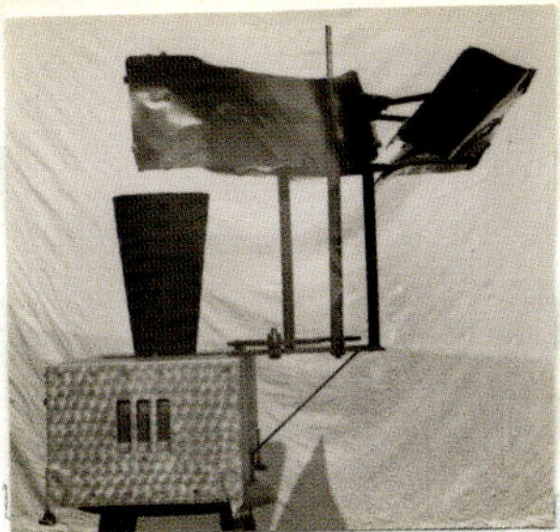

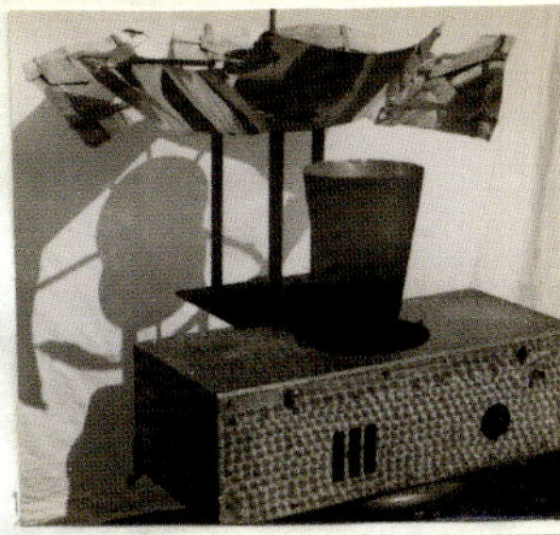

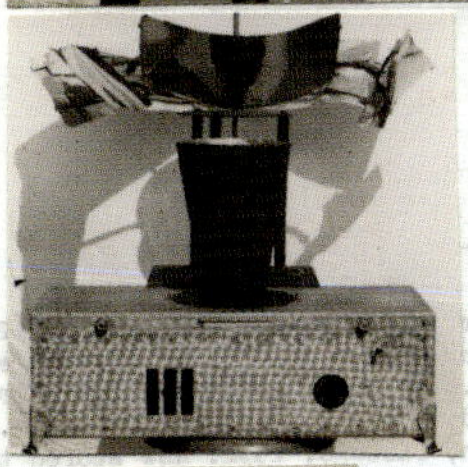

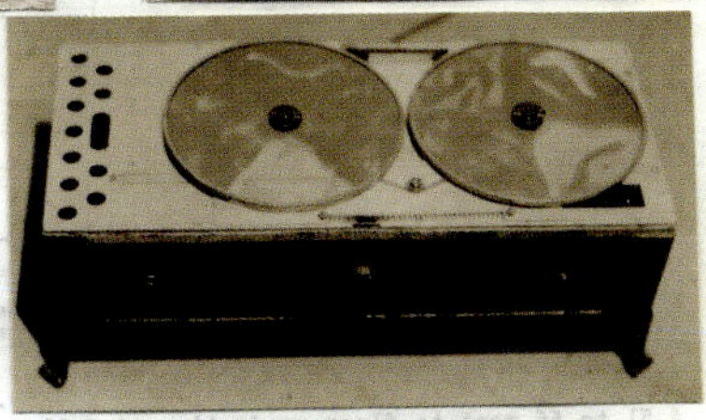

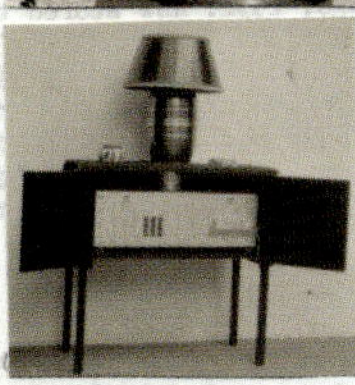

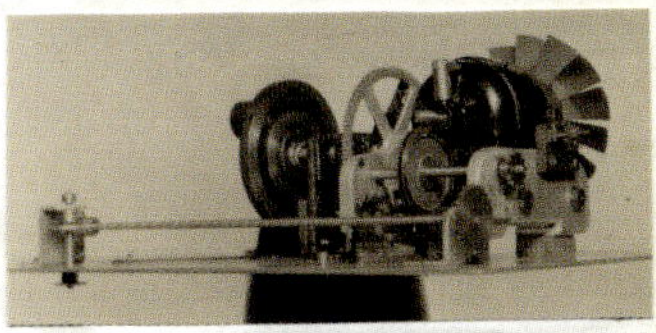

First table model Clavilux, # 36. Built 20 W 22 St. in 1928 and installed in cabinet with legs and lamp on top. In 1933 remodeled at 480 Lex. with reflector rotor and used for Study in Depth, Op. 83, from 1933 to 1942. On exhibition at the Museum of Modern Art in the auditorium in the 15 Americans show (in special cabinet) April to June, 1952. Additional specifications on next page.

Slo speed,
at 1 min lamp rev =
motor 1280
100T worm shaft 12.8
Sprocket shaft 6.4
Vert 30T worm shaft 1 rev in 4 1/2 m
Records 1 rev in 11 1/2 m

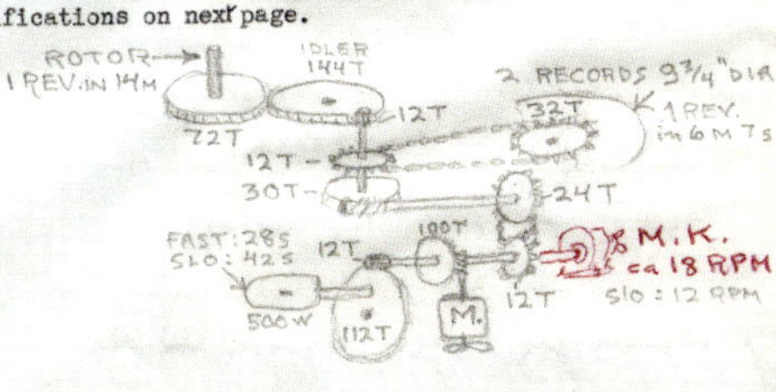

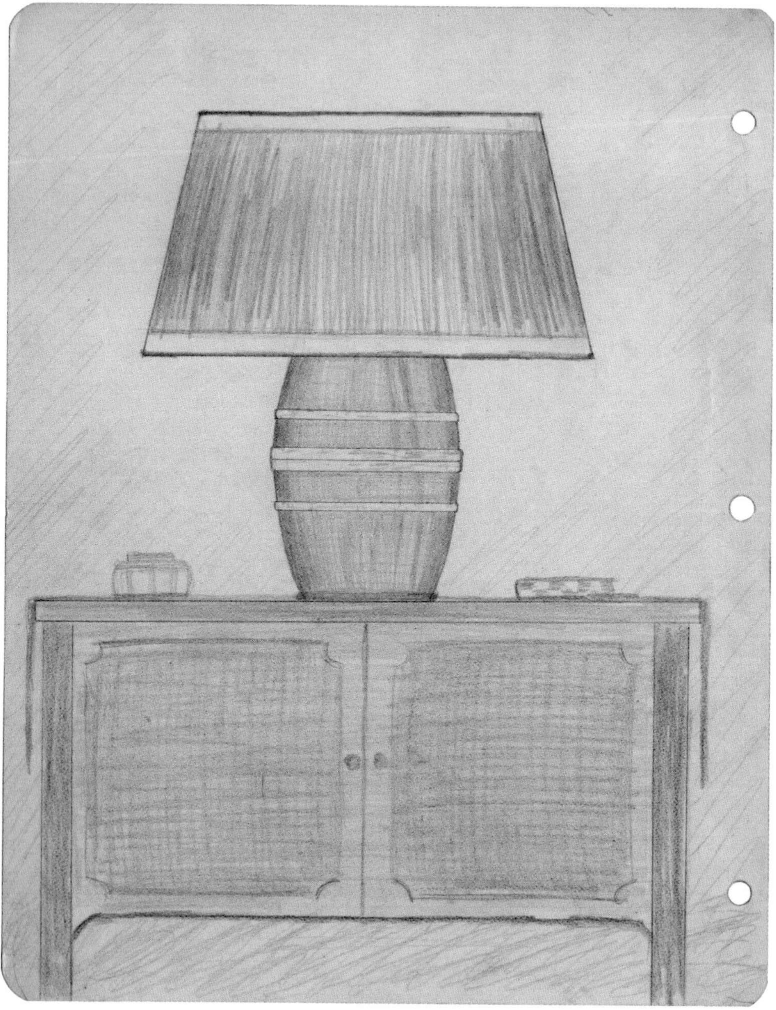

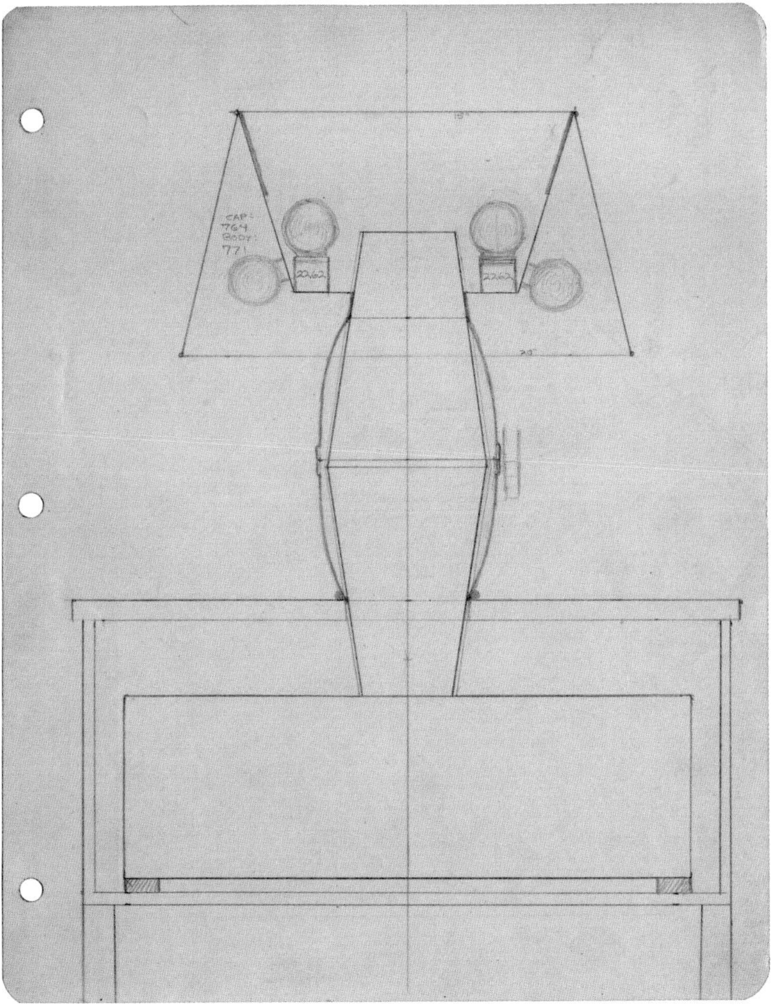
CAP:
764
BODY:
771
2262
2262

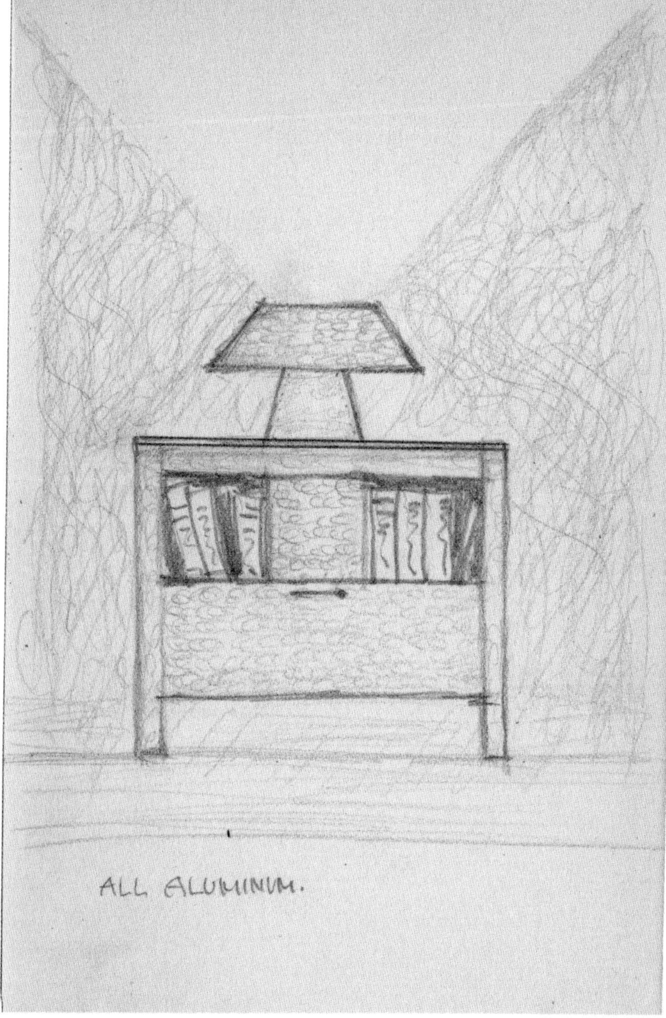

ALL ALUMINUM.

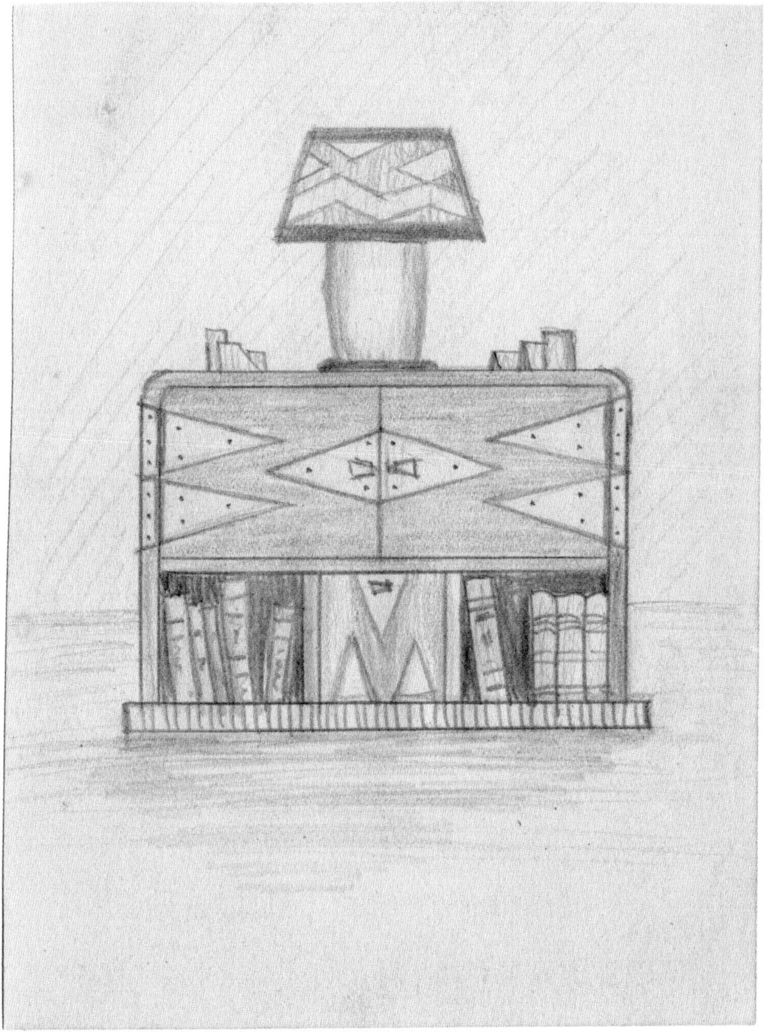

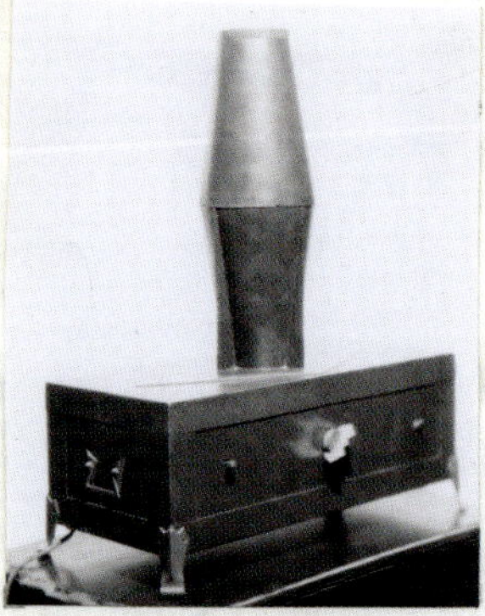

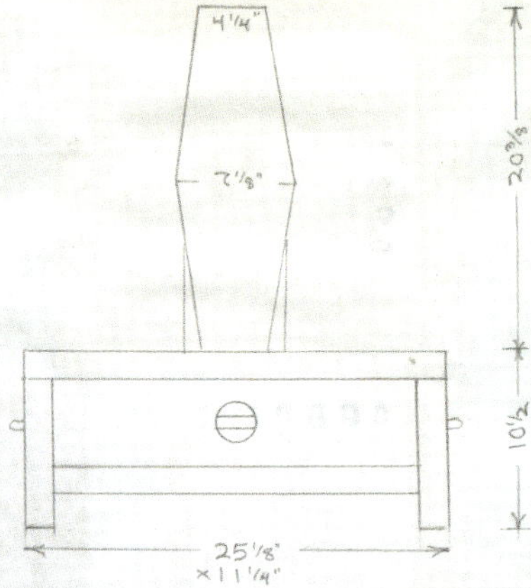

Table clavilux,"luminar" model. Seven of these were built 1928 - 1933 at 20 W 22 St. # 49 to # 55. Rotating 500 w. lamp, two 9 3/4" records on sliding base(see other side.) Double cone internal reflector. Lamp: about 1 RPM. Records: 1 rev in 8 min,30 secs. # 49 went to Dr. Ludwig Kast and later to Mrs. Ladd. # 50 (above) was retained. # 51 to 55 went to floor projection at College Inn, Chicago, 1932. Now stored there.

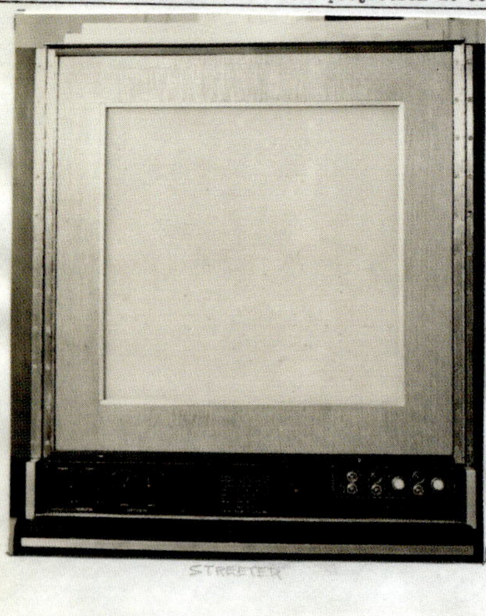

STREETER

FIRST HOME LUMIA INSTRUMENT WITH TRANSLUCENT SCREEN. Five of these were built at 480 Lex. Ave, two with oak, three with aluminum cabinets. 1935 - 36. Turning 150 w lamp in center housing with up and down projection, shutters and color records. Rising forms in upper and selector disc in lower. Adjustable horizon mask. Extension keyboard.

161. Oak. To F.P.B. Washington,1936.
← # 167. Oak.(photo at left) Clevel. Art Mus.
168. Alum. Ansley,Roland, Carnegie Hall.
169. Alum. Alastair Martin,1054.Op.152.
170. Alum. Ralph Houston.

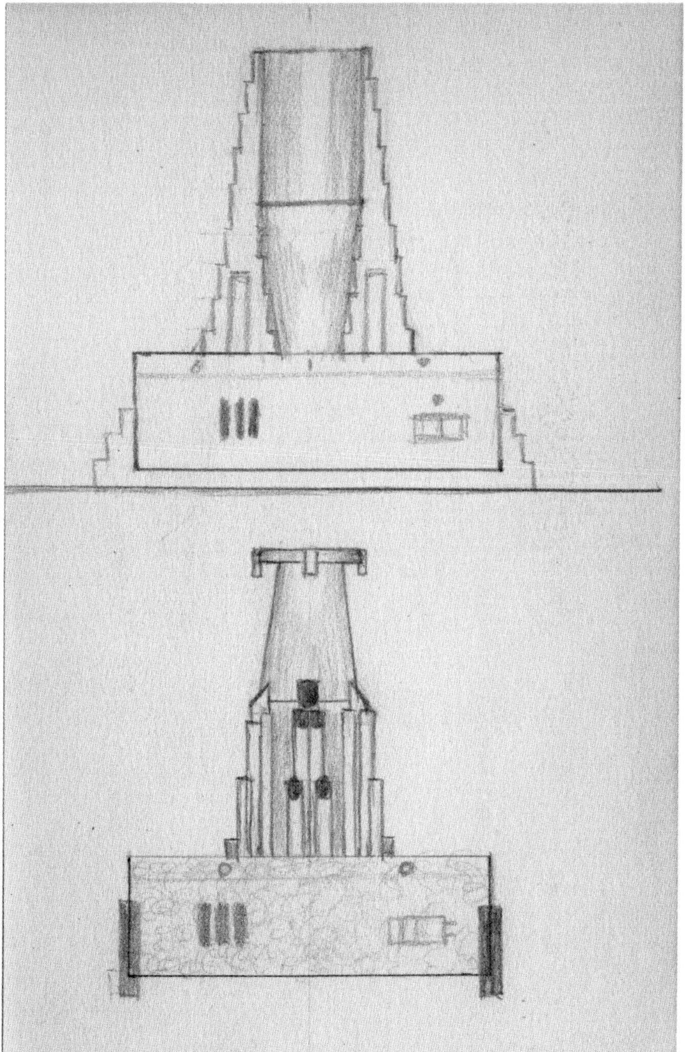

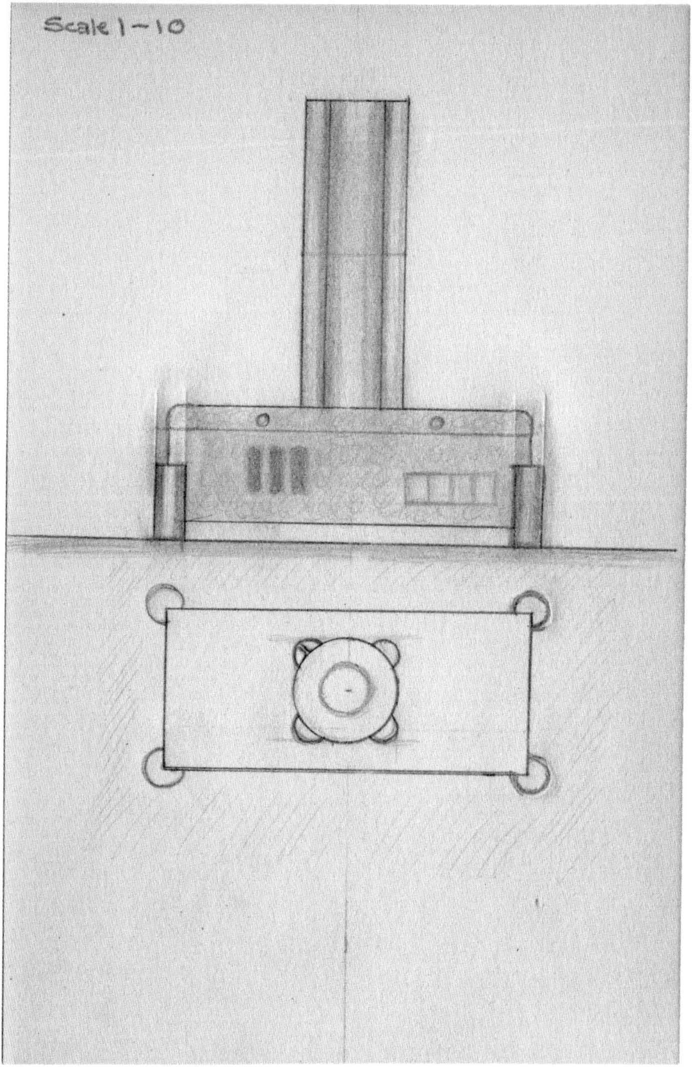
Scale 1-10

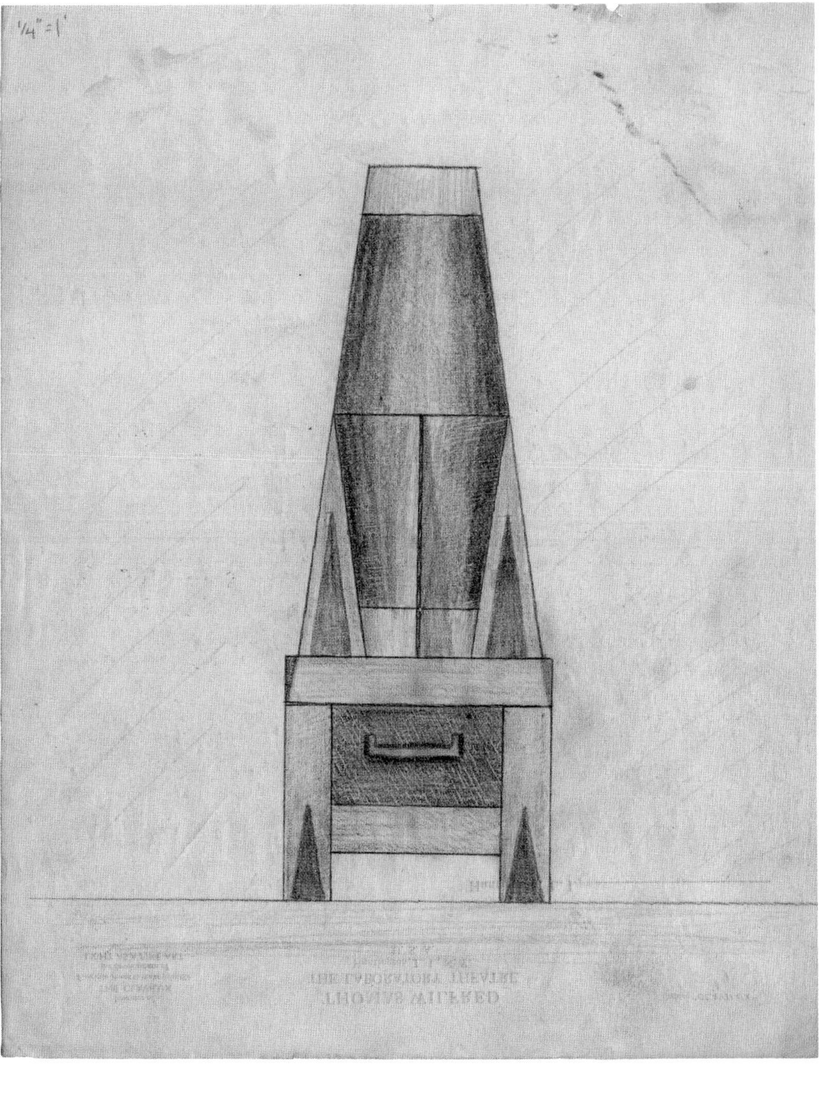
1/4" = 1'

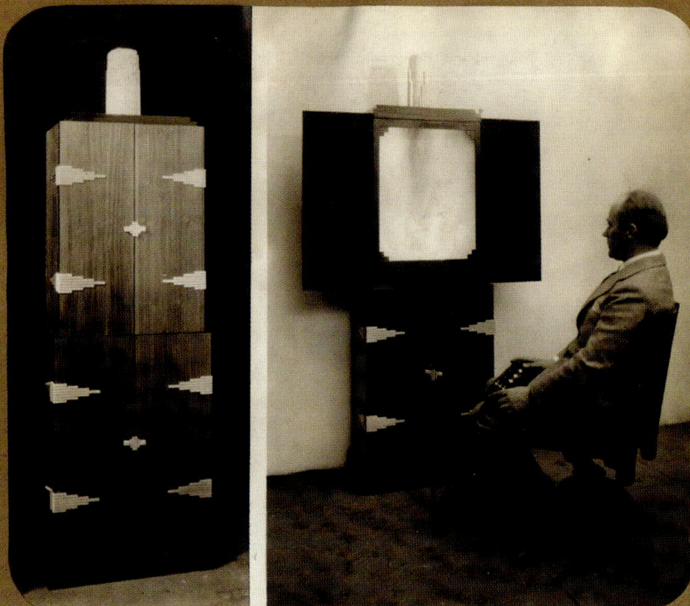

FIRST HOME CLAVILUX MODEL. 16 were built in 1930, 14 in walnut cabinets as above.

\# 82. Small cabinet with round screen. Speiser, 1930. Dead. No record.
83. Colonial cabinet. Joffe. 1930. Dead, No record
84. Walnut, stmaight. Angier. 1930
85. -"- , corner. Frances Lee. 1930
86. -"- , straight. Dr. Magrath. 1930. Later to Mrs Spec '38.
87. -"- , straight. Samuel Kimball. To Grant Waddle Milan Ohio '45
88. -"- , straight. Martin Schwab. at his death to a friend Chicago '34
89. 9"- , straight. Helen Cooke. Dead. No record
90. -"- , straight, no hinges, Katherine Dreier. Dead no record.
91. -"- , straight, spec. A.C. No keyboard. Westinghouse Lighting Institute 30-31. W. Nyack
92. -"- , corner. R.C.A. Camden, Dan Williams, Clyde St. W. Nyack.
93. -"- , corner. Ella Crandall. 1935. Died '38 W. Nyack
94. -"- , straight. Sewell Avery. Dead. No record
95. -"- , straight, no hinges, Stokowski.
96. -"- , straight. Unit to Conway '51. Cabinet at W. Nyack
97. -"- , straight. Lix-Klett, Argentine.

All had 100 w moving lamp, double cone and changeable records stored in base stack. Units 83 to 97 had extension keyboards controlling shutter and three small floods, as well as tempo and stop. Screens were curved opaque. Cabinets were built by furniture firm.

FIRST HOME CLAVILUX

Clavilux Junior series, 1930

"The last of the evening is devoted to the greatest invention in relation to the art of painting which as yet has been achieved. I have asked Mr. Wilfred to bring his home-clavilux so that you can come on stage after the performance and look at it for yourself. For this little instrument will, I hope, eventually enter into every home and develop a love for color as the radio and gramophone have developed a love for great music."

Excerpt from a draft of Katherine Dreier's lecture "The Art of the Future," delivered March 23, 1931, at the New School for Social Research, New York. Wilfred's performance followed films by Lotte Reiniger and Marcel Duchamp and a demonstration by Alexander Archipenko of his *Archipentura*.

Thomas Wilfred continued his attempts to market a domestic scale model with the 1930 Clavilux Junior. *These units came with a selection of hand-painted records that allowed users to choose their own composition, as well as a "remote control" that enabled them to change the tempo, intensity, and brightness. Wilfred's detailed inventory (page 36) of the sixteen Home Clavilux models that he built notes the design of each custom-made cabinet, to whom they were sold, and their current whereabouts.*

Cabinet #90, *described as being "Walnut, straight, no hinges," was acquired by Katherine Dreier, the great patron and promoter of modern art. Dreier was trained as an artist and very excited about the possibility of creating her own colored light works with Wilfred's controllable system. When Dreier ordered her* Clavilux Junior, *she asked that the cabinet not include art deco hinges so that it might look less like furniture.* Cabinet #95 *was sold to famed conductor Leopold Stokowski and* Cabinet #88 *to Frances Glessner Lee, the International Harvester heiress and influential forensic scientist who is best known for the dollhouse crime-scene models she created to train homicide investigators.*

OPPOSITE: Unit #86, *from the* Clavilux Junior *series, 1930. Carol and Eugene Epstein Collection*

PAGES 40–41: *Images of* Unit #86 *with two of several light projection possibilities*

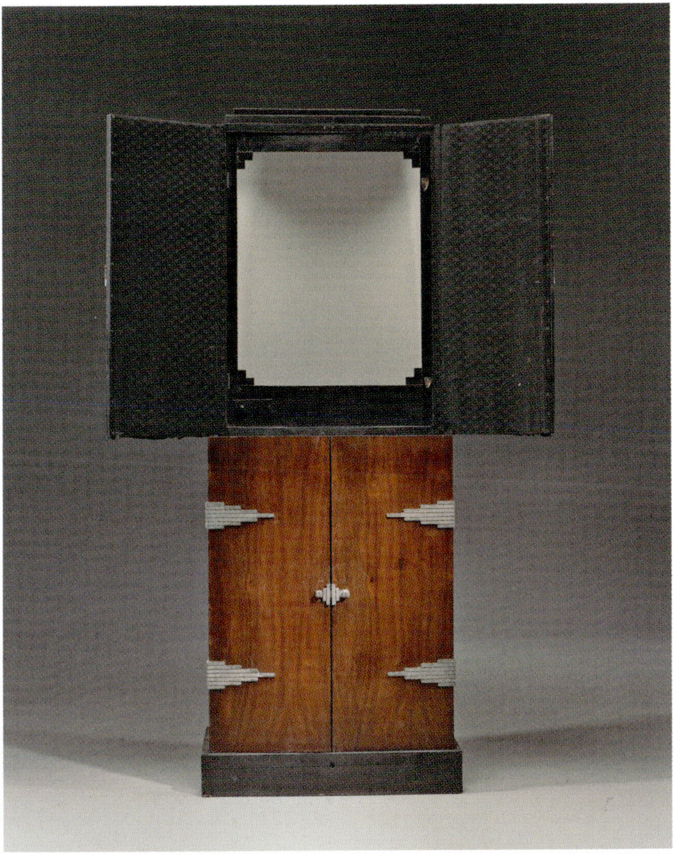

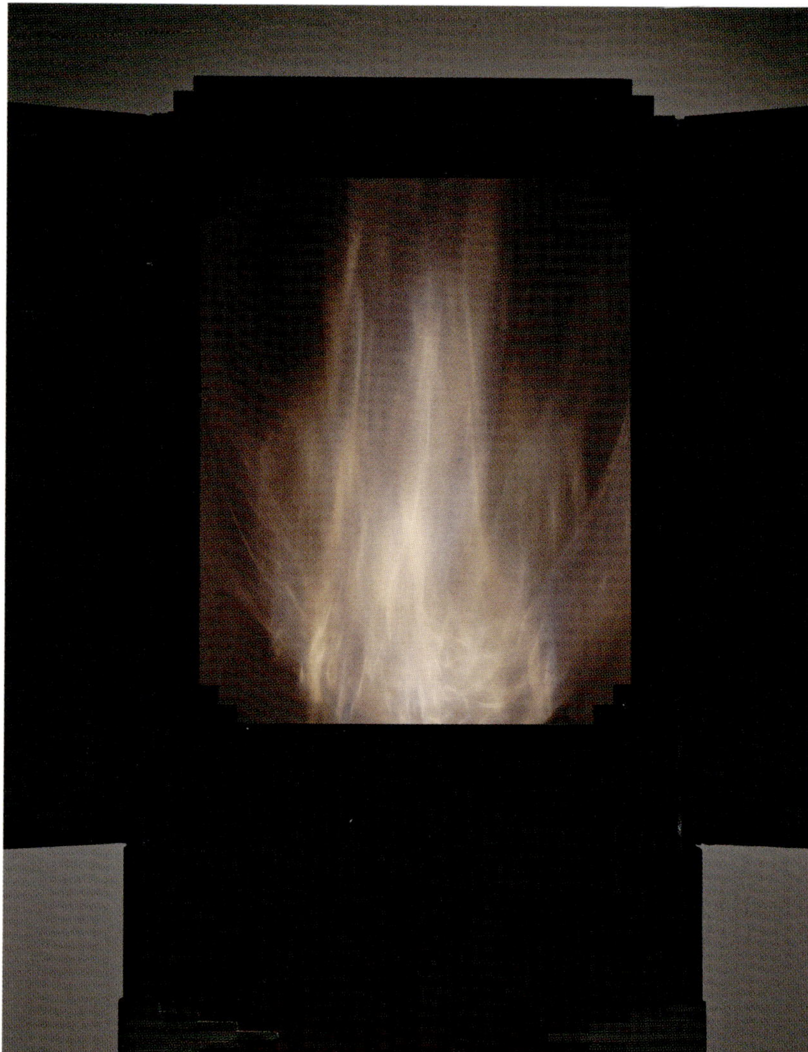

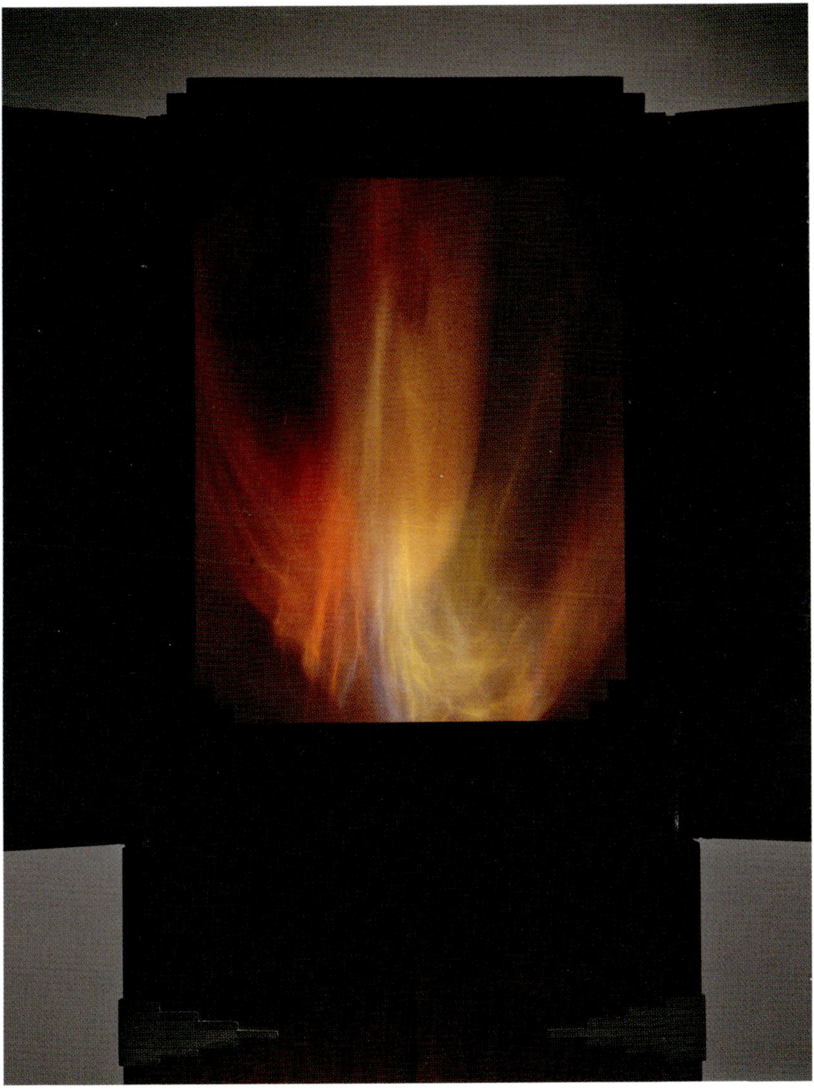

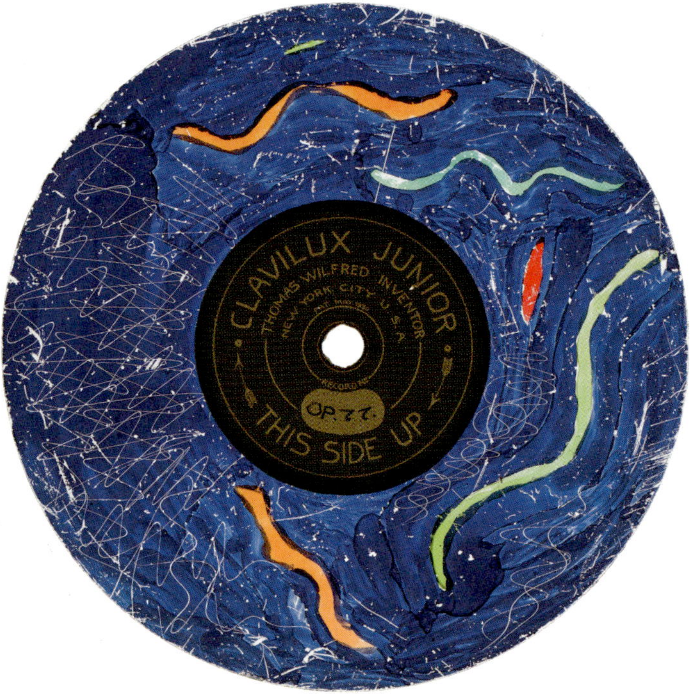
CLAVILUX JUNIOR
THOMAS WILFRED INVENTOR
NEW YORK CITY U.S.A.
RECORD No
OP. 77.
THIS SIDE UP

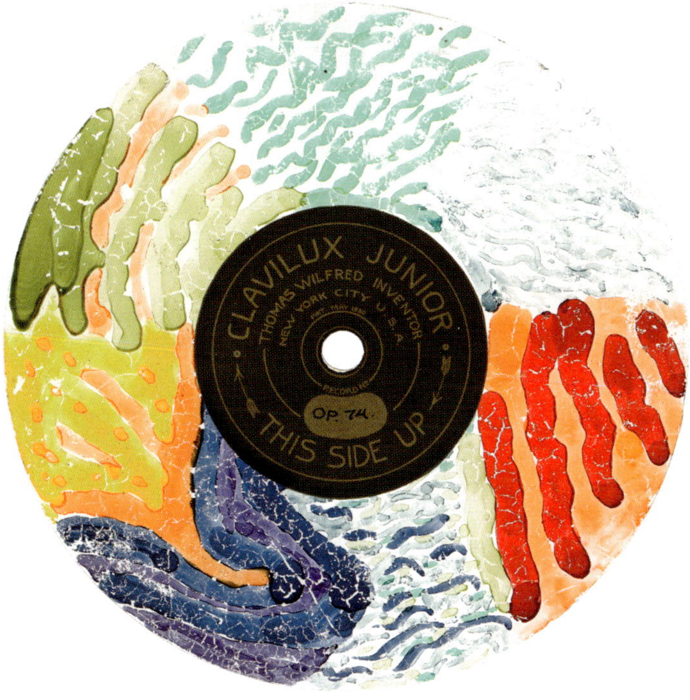
CLAVILUX JUNIOR
THOMAS WILFRED INVENTOR
NEW YORK CITY U.S.A.
Op. 74.
THIS SIDE UP

PAGES 42–43: *Two of the hand-painted glass records that originally accompanied the* Clavilux Junior Unit #86

OPPOSITE: *Wilfred's operating instructions for Clavilux owners. If the Clavilux broke, owners were instructed to "close the instrument and call the Art Institute of Light, Eldorado 5 4788 for service."*

PAGES 46–47: *Thomas Wilfred display at the Electric Show, Madison Square Garden, September 20–30, 1933*

DIRECTIONS FOR THE OPERATION OF CLAVILUX UNIT.

OPENING THE INSTRUMENT:

Unlock and pull down lower flap - then open both doors fully.

AUTOMATIC OPERATION:

Turn on Main Switch on Extension keyboard on right side of lower panel.Turn on All Motion switch next to Main. Press upper or lower Shutter button,or both and wait for image to appear.Set Horizon and Diffusion keys on left side of lower panel until desired sequence has been obtained. The lower flap may then be raised and locked and the instrument left to play the pre-set sequence.

HAND OPERATION:

Pull out Extension Keyboard as far as the cable permits and arrange a chair as far from the instrument as the keyboard reaches.Set Horizon and Diffusion keys as desired and hold keyboard in left hand.Turn on Main Switch and All Motion without which none of the switches or buttons will operate.

SHUTTER BUTTONS:

The two large white buttons release slow moving discs which are alternatingly trans~~lucent~~ parent and opaque.If,therefore,the shutter is closed when the button is pressed it will open,if it is open it will close.In intermediate positions the buttons have no effect - it is necessary to wait for the shutter to open or close completely before pressing again.

VERTICAL AND HORIZONTAL MOTION:

These two movements have both button and switch.The button is for brief movement from one position to another,the switch for continuous motion.

IMPORTANT: IT IS ADVISABLE TO CLOSE AND LOCK THE INSTRUMENT AFTER USE.The left door actuates a second Main Switch within the instrument,shutting off all electric current.If the doors are left open BE SURE TO TURN OFF MAIN SWITCH before leaving as otherwise the lamp may burn,with both shutters closed,for weeks unnoticed.This will damage the color filters..Better CLOSE AND LOCK THE INSTRUMENT EACH TIME AFTER USE !

If instrument does not work when Main and All Motion switches haze been turned on and shutter buttons pressed,close the instrument and call the Art Institute of Light.Eldorado 5 4788 for service. DO NOT ATTEMPT TO OPEN PROJECTION CHAMBER AS THIS WILL VOID GUARANTEE !

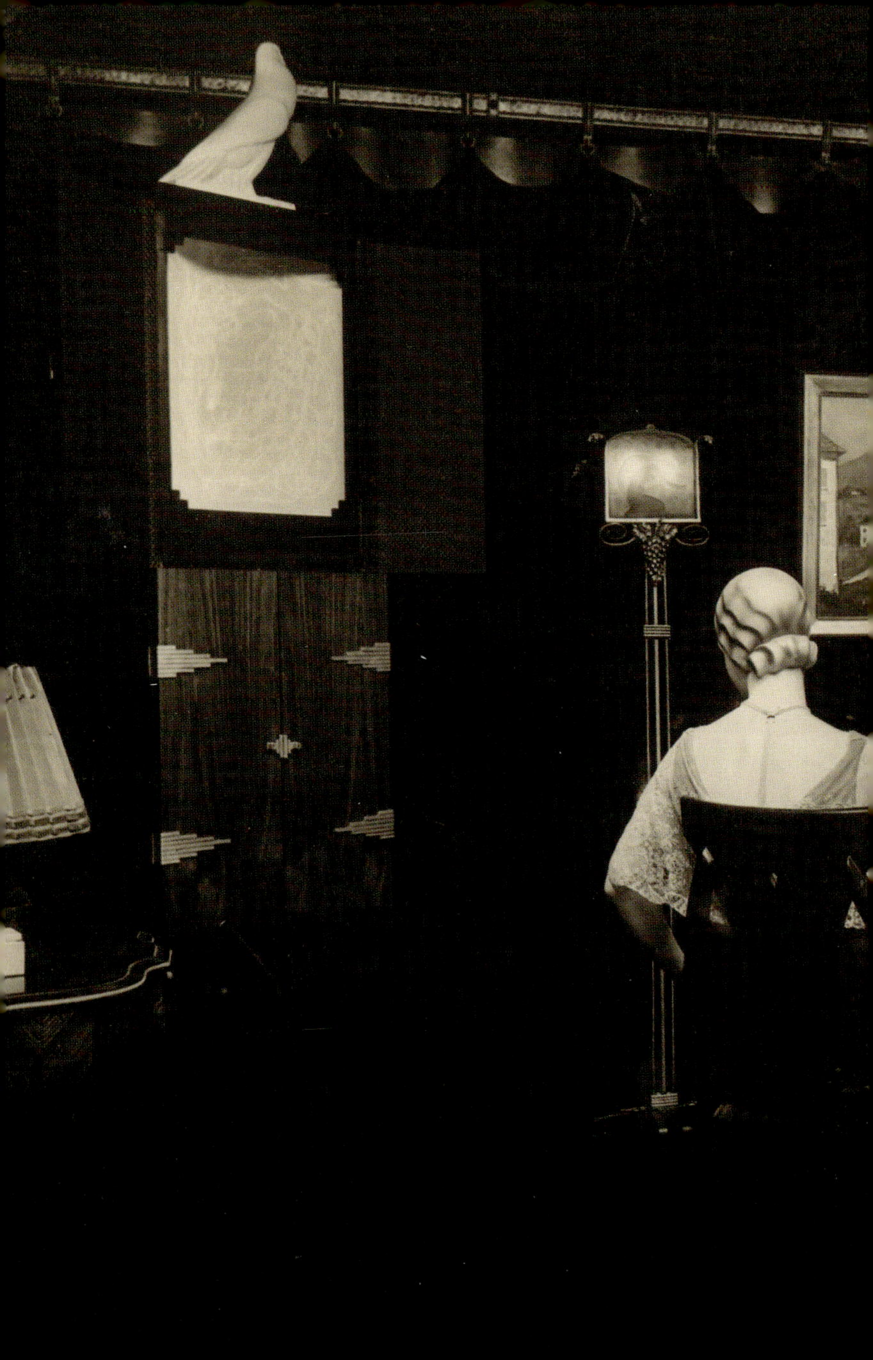

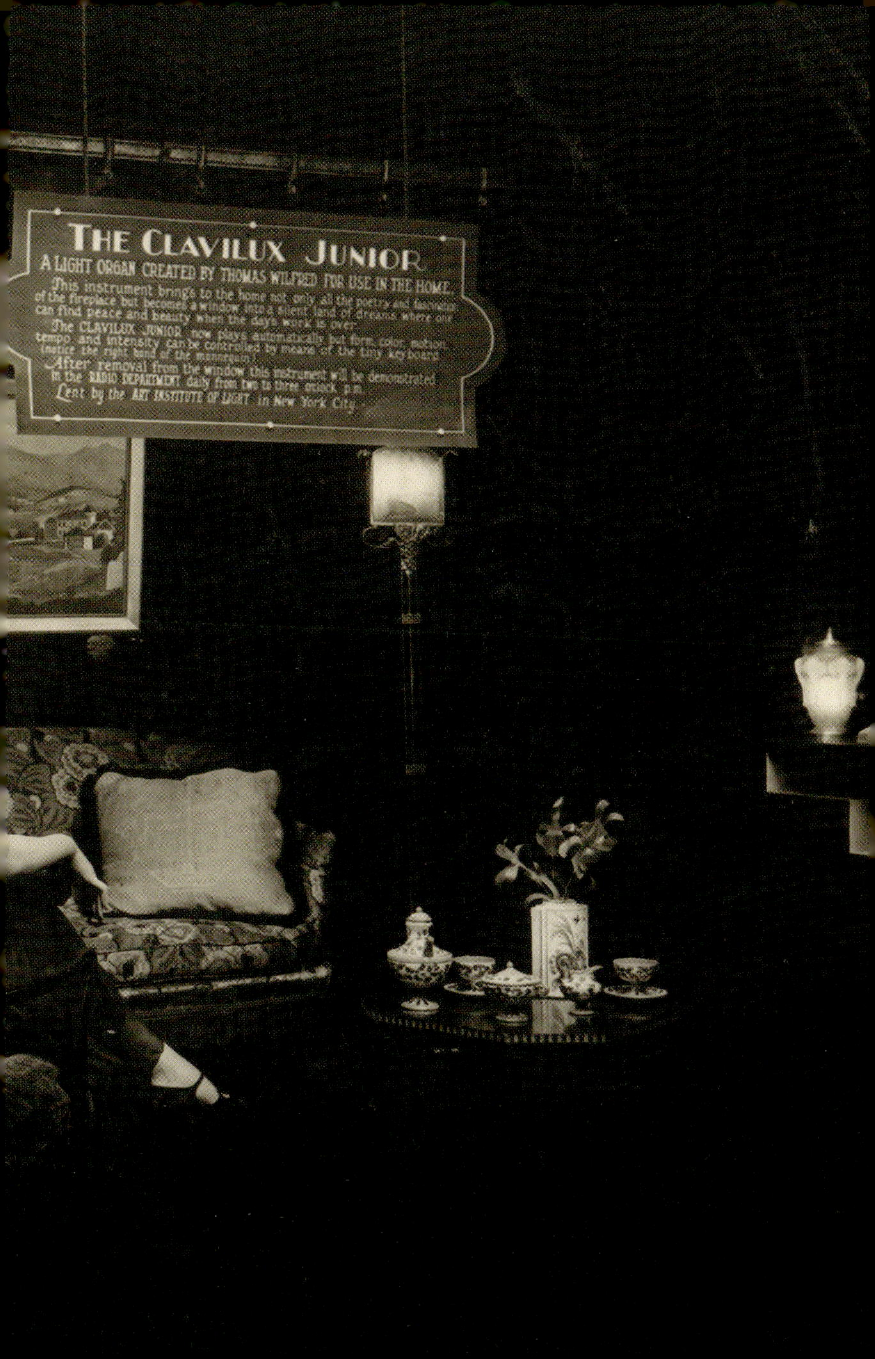
THE CLAVILUX JUNIOR
A LIGHT ORGAN CREATED BY THOMAS WILFRED FOR USE IN THE HOME.
This instrument brings to the home not only all the poetry and fascination
of the fireplace but becomes a window into a silent land of dreams where one
can find peace and beauty when the days work is over.
The CLAVILUX JUNIOR now plays automatically but form, color, motion,
tempo and intensity can be controlled by means of the tiny keyboard
(notice the right hand of the mannequin)
After removal from the window this instrument will be demonstrated
in the RADIO DEPARTMENT daily from two to three o'clock p.m.
Lent by the ART INSTITUTE OF LIGHT in New York City.

Thomas Wilfred with a Clavilux Junior, ca. 1930. Wilfred designed several different cabinets to house these works. The one pictured here could be Clavilux Junior #82, *which is listed on the inventory (page 36) as a "Small cabinet with round screen. Speiser, 1930." In this photograph Wilfred collaged an image from one of his large-scale theatrical projections into the screen area of the cabinet. The bookshelves below and decorative elements atop the unit also feature in his pencil studies for the Luminar and the Clavilux Junior home models.*

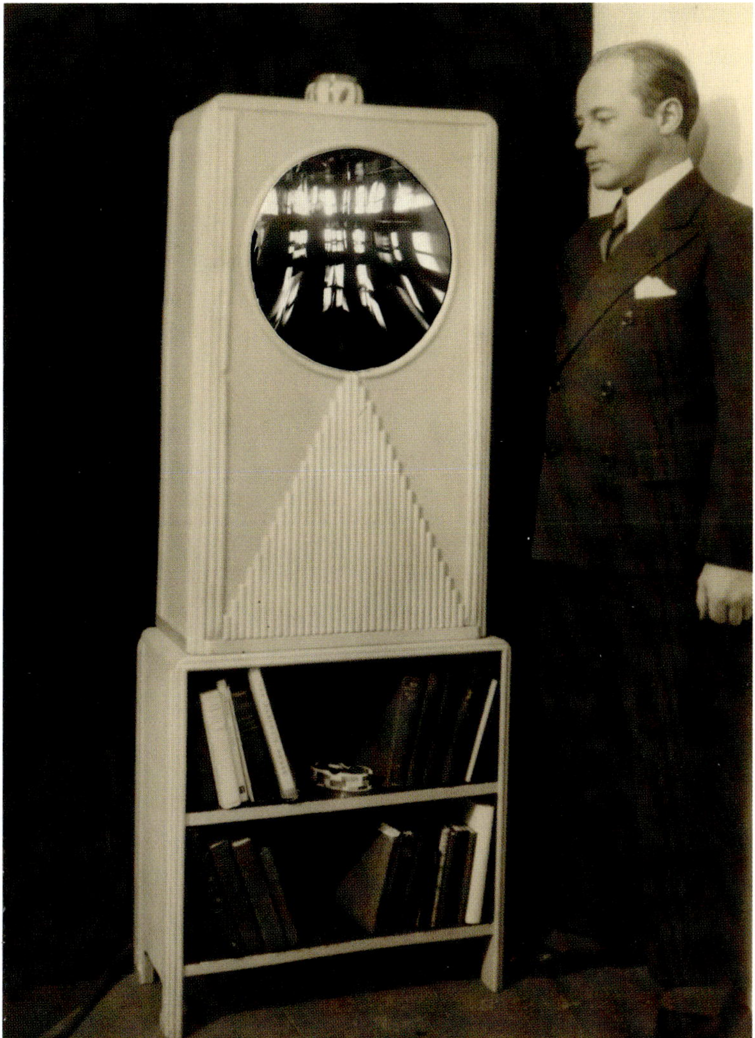

LUMIA HOME MODELS

Tranquil Study, Op. 92, 1935
Multidimensional, Op. 79, 1932

In 1933 Wilfred opened the Art Institute of Light at Grand Central Plaza in Midtown Manhattan. This nonprofit organization, which had more than four hundred members by 1935, was devoted to the promotion of the art of Lumia and included a research center, a technical library, and a large theater where Wilfred demonstrated his Clavilux and Lumia instruments. It was here that Wilfred began to develop a new form of small-scale instruments that no longer had the appearance of furniture. They were designed to be mounted on walls and pedestals, making them more suitable to the public spaces and museums where Wilfred's work was increasingly finding an audience.

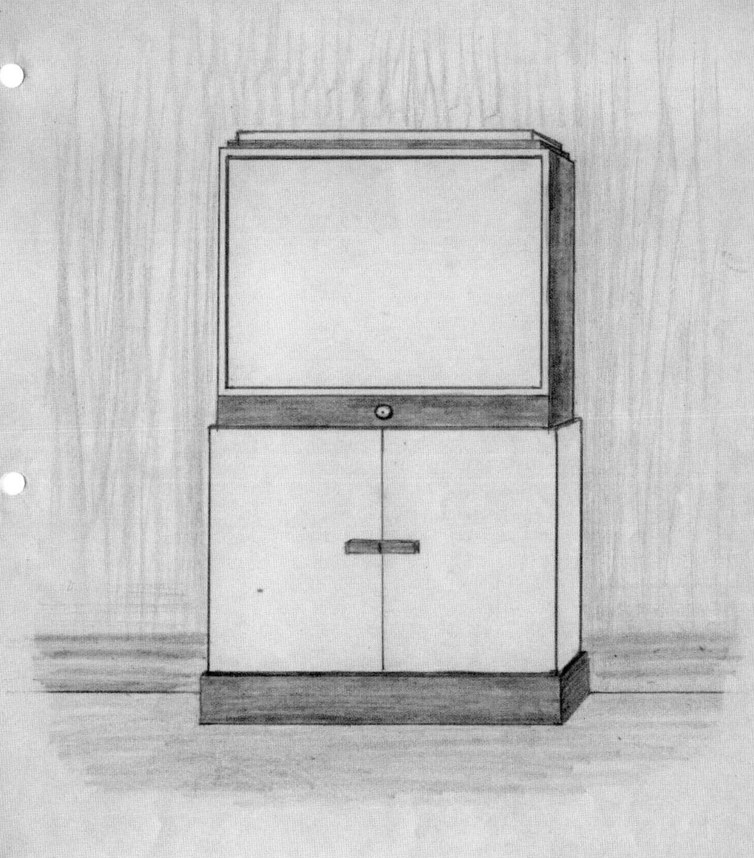

height 6'3"

OP 92. TRANQUIL STUDY
1935 5 M 155

1" = 1'0" Screen 39 x 29 Overall of top section 42 x 36 x 10"

PAGE 51: *Drawing for* Tranquil Study, Op. 92, *1935*

OPPOSITE: Tranquil Study, Op. 92, *1935. Collection Yale University Art Gallery, New Haven, Conn., Gift of Thomas C. Wilfred*

PAGES 54–55: Multidimensional, Op. 79, *1932, and one of the many images that make up Wilfred's "recorded" composition for this work*

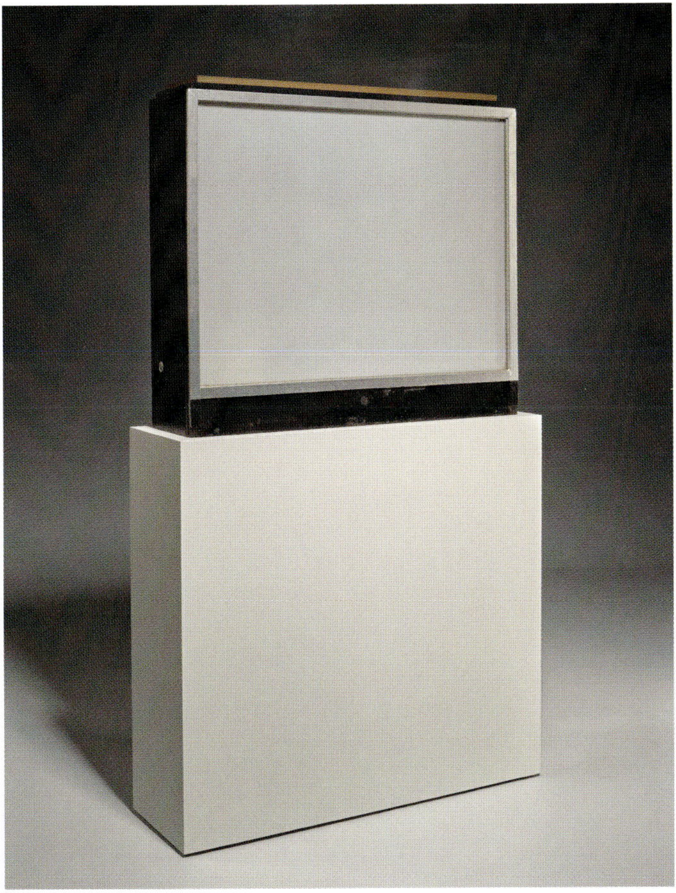

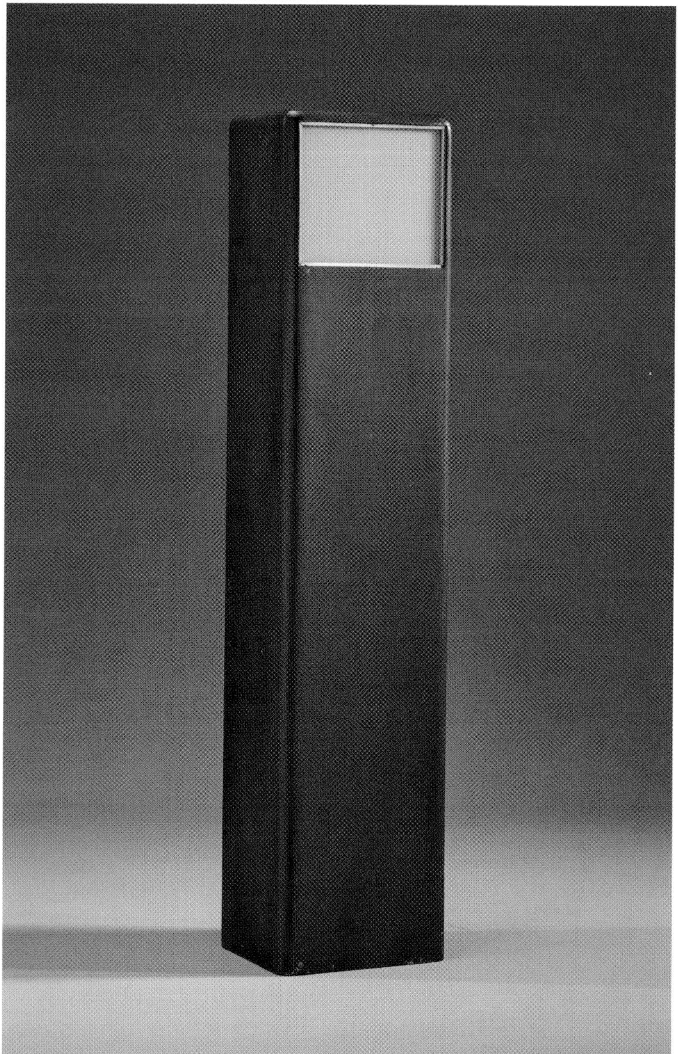

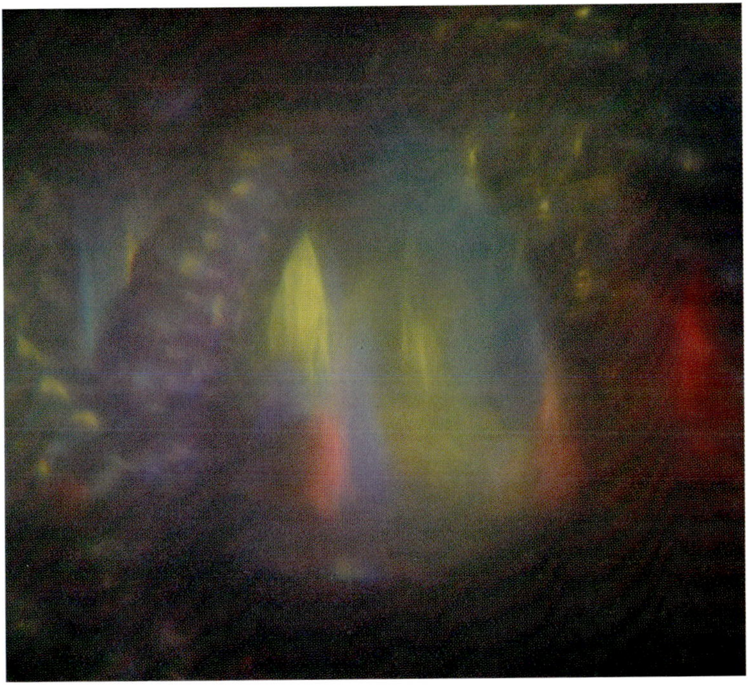

Wilfred's work was included in the 1952 exhibition 15 Americans *at the Museum of Modern Art in New York. Organized by Dorothy Miller, the exhibition also included Frederick Kiesler, Jackson Pollock, and Mark Rothko. Five Wilfred works were shown, including* Multidimensional, Op. 79 *(see page 54).*

PAGES 58–59: *Artist's statement by Thomas Wilfred from the MoMA catalogue for* 15 Americans, *1952*

PAGES 60–61: *Installation plan for Thomas Wilfred's "Lumia Compositions" in the exhibition* 15 Americans, *1952*

THE MUSEUM OF MODERN ART NEW YORK

15 AMERICANS

THOMAS WILFRED

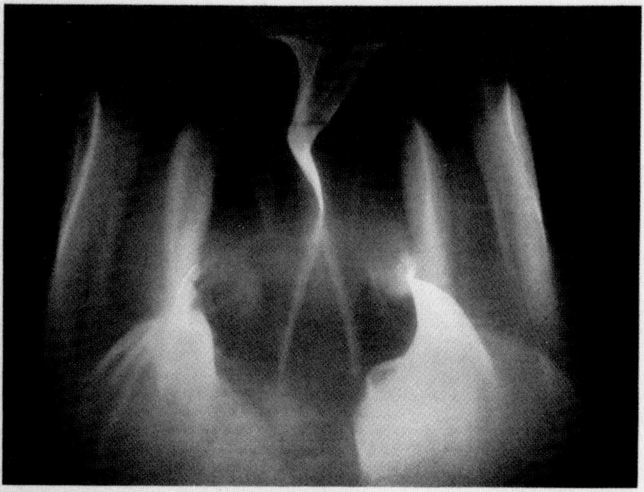

THOMAS WILFRED: A passage from the lumia composition. *Abstract, Opus 59.* 1923

A new art form has begun its life in our generation, a silent visual art in which the sole medium is light. This new art form has been named lumia.

Like the older arts, lumia is an esthetic concept expressed through tools, materials and techniques. The lumia artist visualizes his composition as a drama of moving form and color unfolding in dark space. In order to share his vision with others he must materialize it. This he does by executing it as a two-dimensional sequence projected on a flat white screen by means of a specially constructed light-generating instrument controlled from a keyboard. By manipulating the sliding keys, he can release white light, mold the light into form, add color, and imbue the result with motion and change.

Lumia is neither composed nor performed like music. In music instruments have been evolved, and long since standardized, by means of which any composition may be played instantly. In lumia the execution of nearly every new work means changes in the existing equipment. Lumia may never be played in the manner of music and I see no reason at all for striving toward this goal. The two arts are so different in nature that attempts to design lumia instruments in imitation of musical ones will prove as futile as attempts to write lumia compositions by following the conventional rules laid down for music. We must shun imitation and deal with lumia in terms of itself.

The result alone matters. How we have achieved it is a secondary consideration, whether we have adhered to established procedures in optics, or kicked over the traces to do something which may well make a physicist scoff but which nevertheless yields the desired result. This applies to all phases of the new art; it is obvious that the rules governing static composition and static color harmony in painting do not apply to form and color in motion, nor do the rules of composition in music apply to a visual sequence.

If a lumia composition is stopped at any point, an analysis of the static image may show both form and color out of balance from the painter's point of view. It will have meaning only when viewed as a link between what came before it and what is to follow. Harmony and balance are here a kinetic concept; the composition may not contain a single moment of static symmetry, no matter where stopped.

A lumia composition may be recorded and mechanically rendered, either full screen size or reduced almost to a miniature, by means of playing devices corresponding to the phonograph in music. The compositions in this exhibition are such recordings.

Lumia has been used in various fields; when played from the keyboard, as visual accompaniment to music, the drama and the dance; and in recorded form, as projected mobile mural decoration, as exhibits in museums and private collections, and as an aid in psychiatric diagnosis and psychotherapy.

THOMAS WILFRED

Born Nestved, Denmark, 1889. Made first experiments with light, Copenhagen, 1905, with a cigar box, an incandescent lamp and pieces of colored glass. Continued experiments while studying art and music, Copenhagen, Paris, London, 1905-11. Professional singer and player of twelve-stringed archlute, 1912-19, to support experiments with light as an independent art medium. Served in Danish army, 1914-16. To U.S.A., 1916; continued concert work until 1919, then gave full time to building an instrument, the "clavilux," completed late in 1921. First public "lumia" performance with the clavilux, January 1922, at Neighborhood Playhouse, New York. Clavilux recital tours, 1922-43, United States, Canada, Europe (first European tour, 1925). Soloist with Philadelphia Orchestra under Stokowski, 1926, performing visual setting

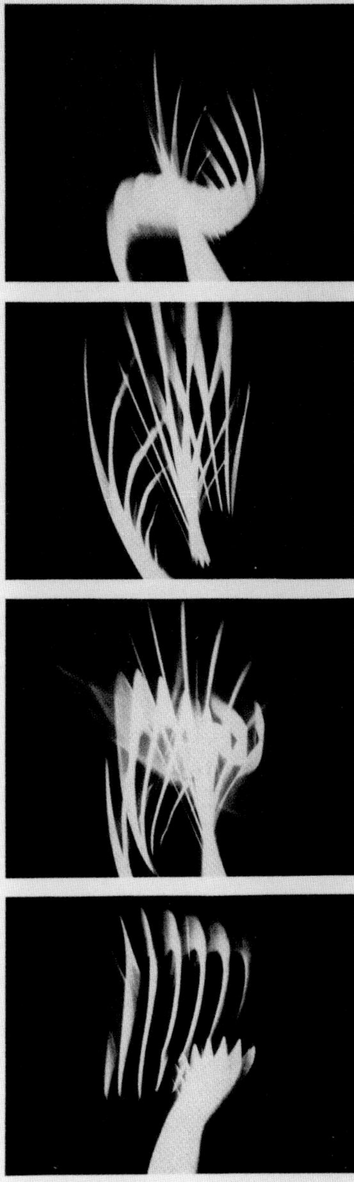

THOMAS WILFRED: A sequence, eight seconds long, from the lumia composition *Unfolding, Opus 127*, 1940

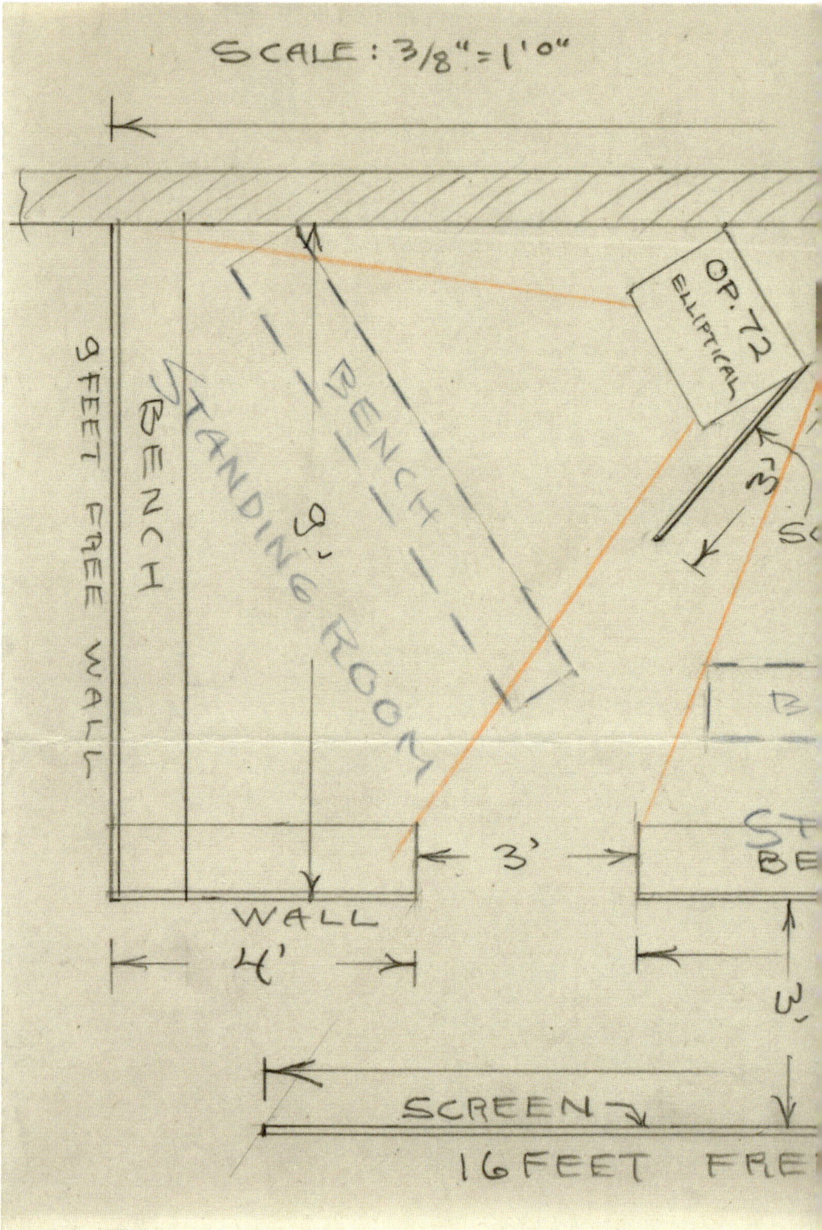
SCALE: 3/8" = 1'0"
9 FEET FREE WALL
BENCH
STANDING ROOM
BENCH
9'
OP. 72
ELLIPTICAL
3'
3'
WALL
4'
3'
SCREEN
16 FEET

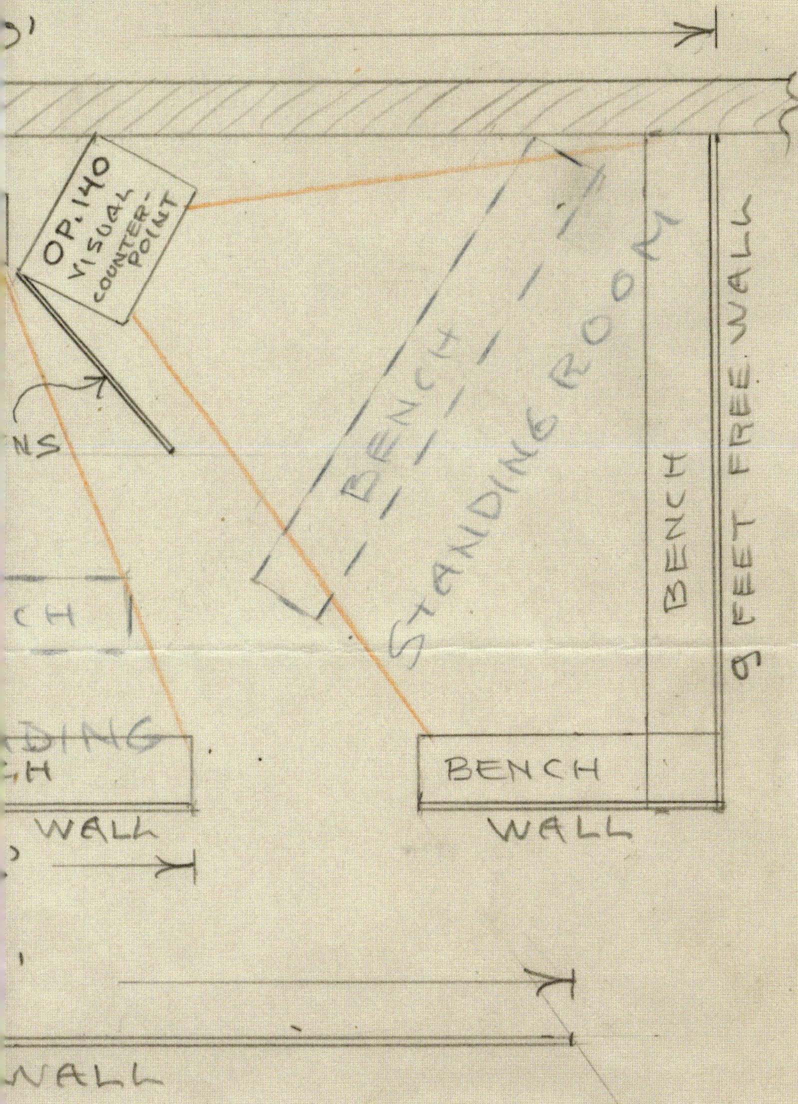

9'
OP. 140
VISUAL
COUNTER-
POINT
NS
BENCH
STANDING ROOM
BENCH
9 FEET FREE WALL
CH
DING
CH
BENCH
WALL
WALL
WALL

THE SOCIETY OF THE NEW YORK HOSPITAL

INCORPORATED 1771

MAINTAINS
THE NEW YORK HOSPITAL
68TH TO 71ST STREETS, YORK AVENUE TO EAST RIVER
THE NEW YORK HOSPITAL — WESTCHESTER DIVISION
WHITE PLAINS, N. Y.

THE NEW YORK HOSPITAL
525 EAST 68TH STREET
NEW YORK, N. Y.

TELEPHONE
REGENT 4-6000

September 21, 1939

Mr. Thomas Wilfred, President
Art Institute of Light
Grand Central Palace
New York, New York

Dear Mr. Wilfred:

As the Institute has requested I am acknowledging the letter written by you and the individual terms of the lease for the instrument which you have built to my specifications for use in the study of fantasy.

The sum of Two Hundred and Fifty Dollars is to be paid from the budget of the Laboratory of Experimental Psychology; and this sum is to cover in full the lease and maintenance of the instrument (Fantascope) for one year, beginning September 8, 1939, the day on which the instrument was delivered by you to the Clinic.

It is agreed that after the above one year period the lease may, at my option, be extended yearly at the nominal rate of $50 per year and as long as the instrument is being used by me in the research project for which it was designed, in view of the conditions of the grant the Art Institute of Light has made for this project.

Because of the narrowness of the door leading to the room in which the Fantascope will be used, it has been necessary to devise other means of getting it in. I am informed that the construction of the steel door frame renders it inadvisable to tear it out. On the other hand, an opening can be made in the wall, from the outside corridor, and the instrument brought in through it. The wall will then be closed up in such a way that should it be desirable

THE FANTASCOPE

1939

"Some years ago, Dr. Norman Cameron, then associate professor of psychology at the Payne Whitney Psychiatric Clinic (Cornell University Medical College) in New York City, came to the Art Institute of Light in search of an instrument which could produce a recorded and exactly timed sequence of form, color, and motion for use in the study of schizophrenia. The project interested me so much that we set aside sufficient funds for the design and construction of an instrument to Dr. Cameron's specifications—a self-contained cabinet with a ground-glass screen in one end and an operating keyboard in the other, from which a complex visual sequence could be produced on the screen with split-second accuracy. We named the unit FANTASCOPE, and it was installed in one of the clinical laboratories, with its screen against a safety-glass window in the wall. In the adjoining room, this window was framed like a picture and here the psychiatrist would bring his patient. The Fantascope was synchronized with a sound recorder, and each remark and exclamation could thus later be identified with a definite visual development on the screen. The procedure may therefore be termed a mobile Rorschach test."

—Thomas Wilfred, excerpt from
"The Fantascope: Lumia in Medicine"

THE FANTASCOPE

LUMIA IN MEDICINE

(Excerpt from LUMIA, THE ART OF LIGHT by Thomas Wilfred. In preparation)

The application of lumia in medicine does not really come within the scope of this book, but is important enough to be mentioned briefly. Some years ago, Dr. Norman Cameron, then associate professor of psychology at the Payne Whitney Psychiatric Clinic (Cornell University Medical College) in New York City, came to the Art Institute of Light in search of an instrument which could produce a recorded and exactly timed sequence of form, color, and motion for use in the study of schizophrenia. The project interested me so much that we set aside sufficient funds for the design and construction of an instrument to Dr. Cameron's specifications - a self-contained cabinet with a ground-glass screen in one end and an operating keyboard in the other, from which a complex visual sequence could be produced on the screen with split-second accuracy. We named the unit FANTASCOPE, and it was installed in one of the clinical laboratories, with its screen against a safety-glass window in the wall. In the adjoining room, this window was framed like a picture and here the psychiatrist would bring his patient. The Fantascope was synchronized with a sound recorder, and each remark and exclamation could thus later be identified with a definite visual development on the screen. The procedure may therefore be termed a mobile Rorschach test. When Dr. Cameron became chairman of the Department of Psychology at the University of Wisconsin, the Fantascope was installed there, in a specially built sound-proof suite, and the work was continued.

Dr. Cameron has kindly furnished the following statement for this book:

"There can be no doubt that lumia presented in a fixed sequence under standardized conditions will one day yield research results comparable to those of the Rorschach test of imagination. The difficulties in standardization are greater than in the case of the Rorschach because of the kinetic character of the complex visual sequence on the screen, and the difficulties inherent in relating the subject's responses to the exact point in the sequence which was chiefly responsible for calling them out. However, the greater range and the far greater use of color should make it eventually a most important research instrument. Further work with the Fantascope in the area of psychopathology will depend upon the availability of patients and normal controls in a setting that allows for synchronous recording of the patient's responses to the lumia sequences.

"The possibilities of using lumia in therapy have not as yet been thoroughly explored. Incidental observations made in the course of preliminary research studies seem to indicate that the Fantascope may have the same future that music therapy has. Many of the problems of evaluation in this medium are essentially the same as those working in music with patients. One of the advantages of lumia in this area is that it requires visual attention and that it has different dimensions than auditory sequences. For example, there is less simultaneous fusion and there is far more opportunity to study projection in relation to form and color in lumia. It is to be hoped that psychiatrists, psychopathologists and occupational therapists will seek the help and the sympathetic cooperation of the Art Institute of Light in relation to research and therapeutic problems upon which lumia may shed its light."

(Dr. Cameron is the author of THE PSYCHOLOGY OF BEHAVIOR DISORDERS, a Biosocial Interpretation, Houghton Mifflin, 1947.)

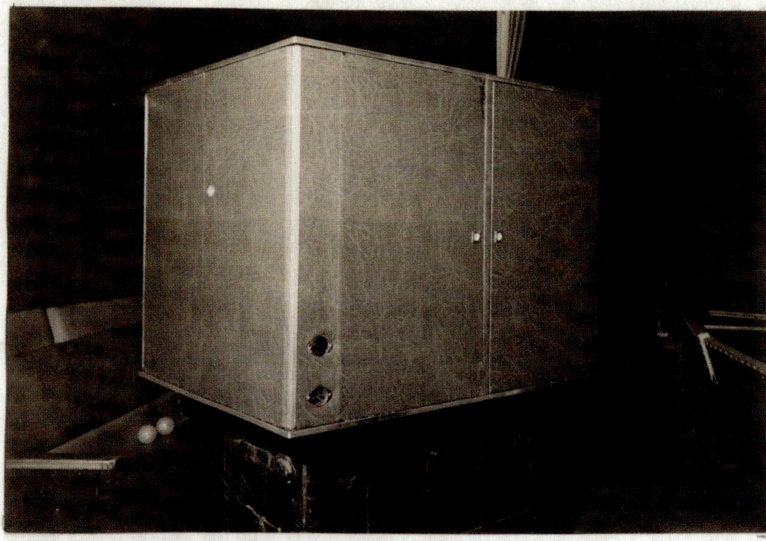

CABINET. CLOSED.

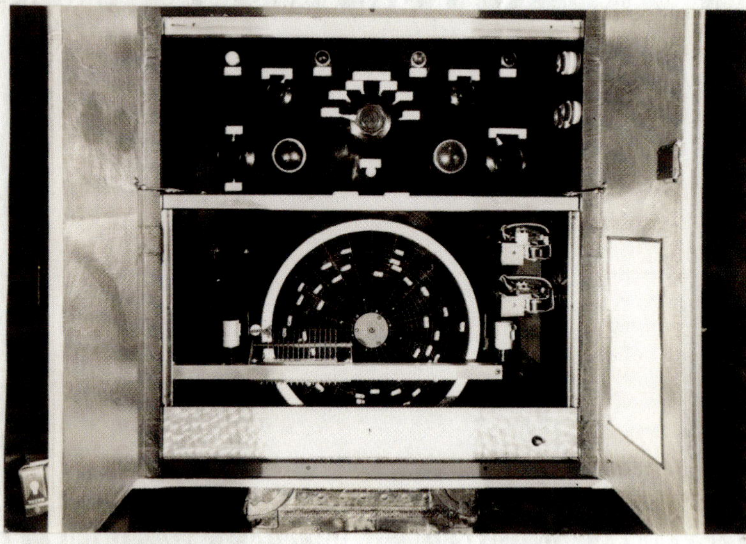

REAR OF CABINET SHOWING CONTROL PANEL AND TIMING COMMUTATOR.

T H E   F A N T A S C O P E

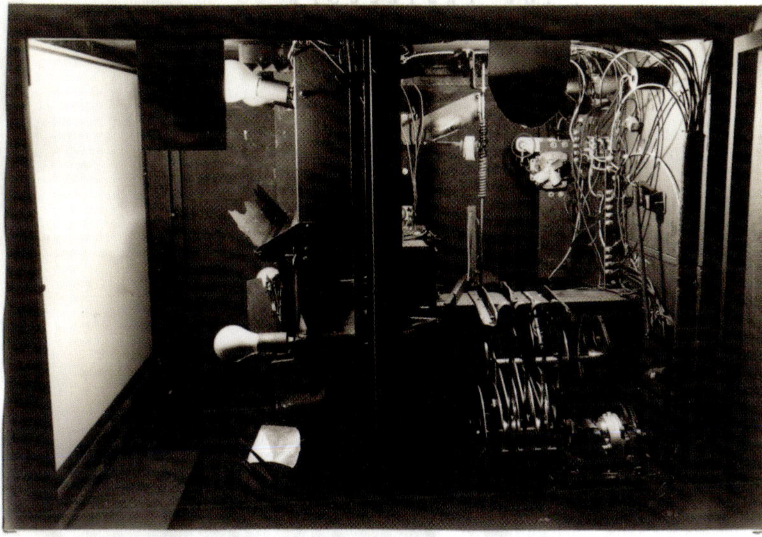

INTERIOR.   RIGHT SIDE.

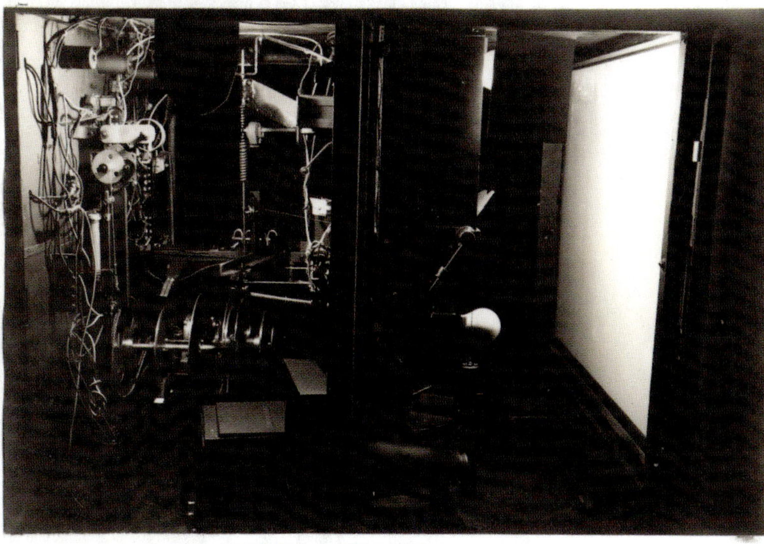

INTERIOR.   LEFT SIDE.

PAGES 66–67: *Excerpt from Wilfred's photographic documentation of* The Fantascope

OPPOSITE: *Programmatic diagram for the operation of* The Fantascope

PAGES 70–71: *Diagram of* The Fantascope *control panel*

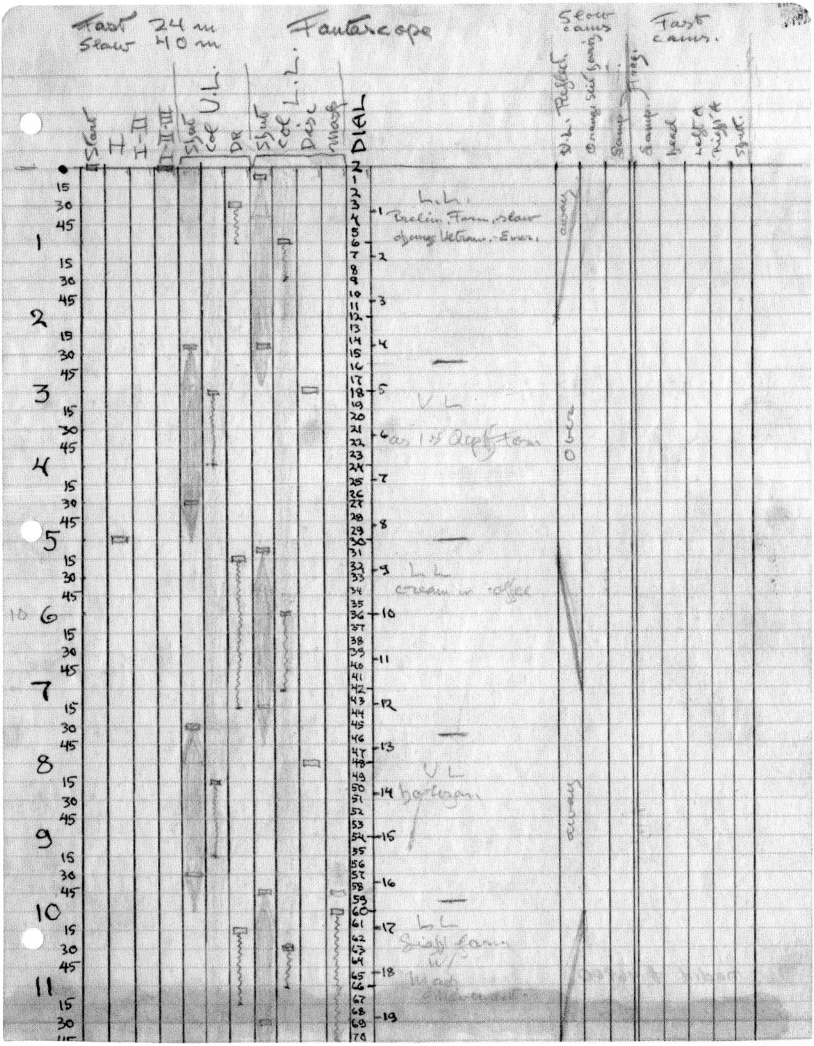
Fast 24 m
Slow 40 m
Fantascope
Slow cams
Fast cams
Slow
I
I-II
I-II-III
Shut col U.L.
DR
Shut col L.L.
Disc
Mask
DIAL
U.L. Reflect
Lamp
Lamp
head
left & right
shut
L.L.
U.L.
L.L.
cream in coffee
U.L.
horizon
L.L.
away
Over
away

T H E F A

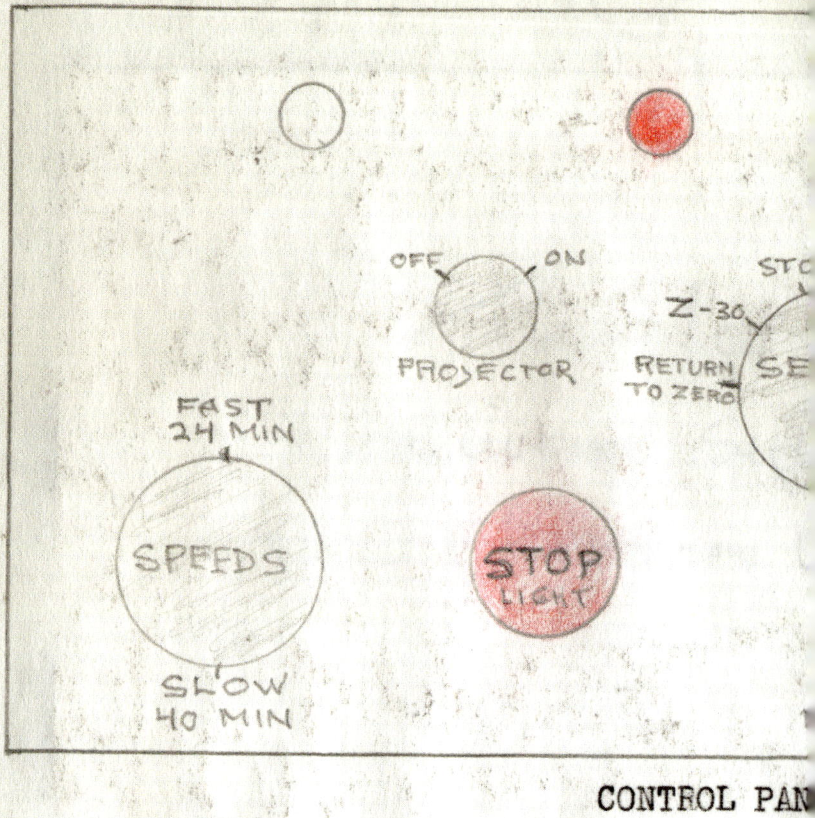

CONTROL PAN

T A S C O P E

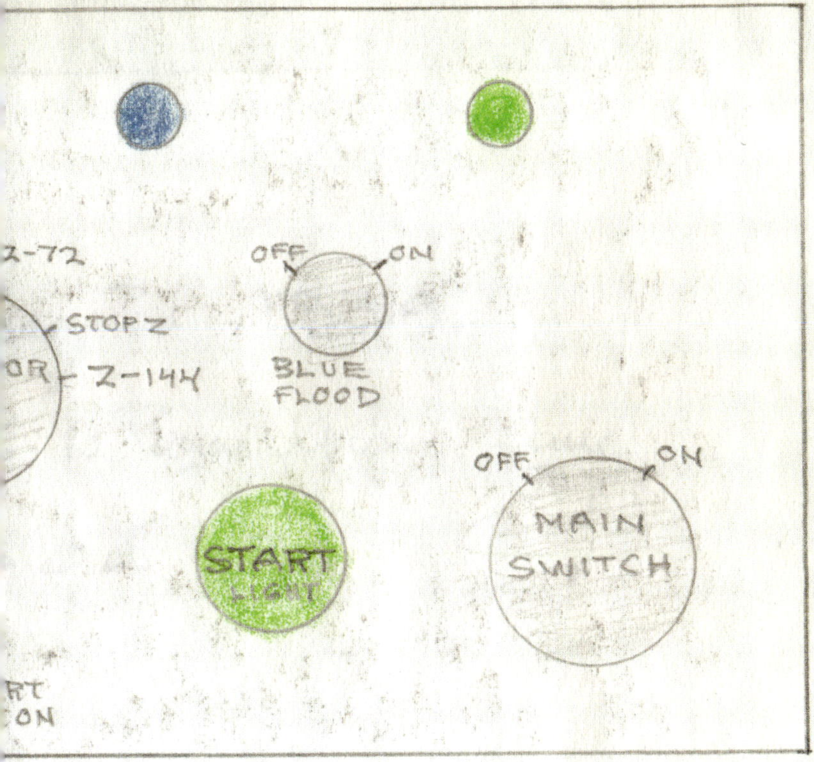
2-72
STOP 2
OR
2-144
OFF
ON
BLUE
FLOOD
START
LIGHT
OFF
ON
MAIN
SWITCH
RT
ON

USE OF COLOR MUSIC IN 'THE VIKINGS' FORECASTS GREAT SCENIC POSSIBILITIES FOR FUTURE

Color Organ to Solve Problems of Ibsen Play

Invention Has Uses in Other Fields

By GILBERT SWAN
NEA Service Writer

NEW YORK, May 17.—For many a year Ibsen's "The Vikings" has baffled the theater producer because it required background visions which could not be put upon the stage.

Then along came Thomas Wilfred with his Clavilux, better known as the color organ, which could produce not only color music but weird abstract color designs as well.

And so "The Vikings" comes to the Broadway belt for the first time and a revolutionary type of scenery also arrives, for the clouds and valleys and peaks of color form which the invention can produce are seemingly limitless and forecast a scenic possibility for the future which may cause the designers a few moments of worry.

With Blanche Yurka leading the cast, this Ibsen fantasy becomes the green play that it is thanks partially to the acting and thanks partially to the unusual scene innovation.

The color organ has, furthermore, other uses—as for instance:

If you have a little spectrum in your home, playing the silent harmonic rhythms of abstract color movement, all emotional disturbances should be swept away, in the belief of Thomas Wilfred, who has adapted his famous "color organ" to home use.

His newest improvisation on the "color music" idea is an instrument ranging in size from an average small radio, which could easily fit in a Greenwich Village hall bedroom, to a parlor sized machine. It uses "color records" as a phonograph plays sound records and the exotic [illegible] of the rainbow [illegible] are observable upon a screen which resembles the sounding board of loudspeaker space of a radio.

One now sits and looks at symphonies and concertos, rather than listening to them.

"It has the dream-like, reverie quality of the old fireplace into which driftwood had been tossed, except that designs and restful pictures are created," says the inventor, who attracted nation-wide attention a few years ago by giving recitals on his larger color organ.

In his Institute of Light, Wilfred admitted, he has been carrying on research in the treatment of psychic disturbances with considerable success.

"The cases I have observed," he reports, "were those brought about by nervous disorders. Extremely [illegible] subjects have been calmed. It has produced a marked relaxation in cases where great tension was noted. While this instrument was not conceived with any notion of [illegible] in psycho-therapy, still the [illegible] in such cases intrigued me and I made a series of experiments. Mind you, I am not claiming any miracles for it. The point is this, if disharmony is at the base of many psychic disorders, then the production of harmony should be sought. And anyone who observes a single record can see at a glance how this result is obtained.

"Of course, at the moment we shall have to select plays to fit the color designs produced in the scenery," he explained. "The Ibsen play has that eerie and fantastic quality which makes the color music background possible.

"It also is being applied to decorative schemes. A big Chicago hotel has just installed one in a special room where the guest can literally choose the color of his own wallpaper."

Thus if you care for a forest-like series of arrangements or a [illegible] vista of colonnades it is necessary only to change the [illegible] color arrangements.

British Parliament Has Pencil Variety

LONDON—(UP)—Styles of clothes, mannerisms of speech and attitudes in debate of members of parliament have all elicited comment by observing members of the house. The latest comment has been about the [illegible] as [illegible] of Front Bench pencils.

Blanche Yurka in Ibsen Play

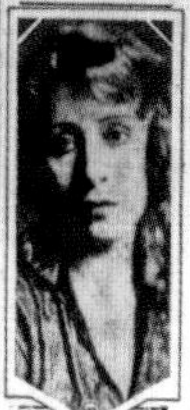

Blanche Yurka's acting will have strong scenic support in the color music to be used in the production of Ibsen's "The Vikings."

Widely Known Orchestra Will Open Coliseum

Complete Decoration by Tuesday

In order to give Tulsa dancers the finest in modern dance music, Bill Grant the manager of the Coliseum Ballroom is bringing the famous Joie Lichter orchestra here, for a two weeks stay beginning next Tuesday evening, direct from a six months' winter engagement at the famous Eagles Ballroom in Milwaukee. Lichter and his unit known as the Merry Mad Gang are all Wisconsin boys but they have furnished entertainment for thousands of dancers in all the big cities of the east during the past three years. Right now they are heralded as one of the best large ballroom orchestras in America.

Work is going ahead at a fast rate on the cooling system and decoration of the auditorium which will be completed by Tuesday evening. A local decorator is in charge of the work and the ballroom will be a thing of beauty when the work is completed.

Sorry He Refused Former Sentence

ST. LOUIS—(UP)—If John A. Piper ever is offered anything again he undoubtedly will "take it."

John was convicted on charges of an automobile theft under the habitual criminal act. He was sentenced to 25 years in prison.

He previously had refused the invitation to plead guilty and be sentenced to two years in jail.

Lois Barall to Head Cast in College Drama

Dramatic Fraternity Sponsors Play

Lois Kahn Barall, recently returned from professional training in dramatics at New York City, has been chosen to play the lead role in the University of Tulsa commencement play, "Adam and Eva," to be presented in the university auditorium Thursday and Friday nights, May 29 and 30, according to Miss Della B. Crowder, director.

Mrs. Barall is an alumnus of Theta Alpha Phi, dramatic fraternity at the university. The fraternity is sponsoring the play jointly with the Faculty Women's club, an organization of women teachers and wives of men faculty members. The cast of the play is to be chosen from active and alumni members of Theta Alpha Phi, Miss Crowder said.

The two organizations are sponsoring it as a project to obtain funds with which to aid Miss Margaret Wyndham, head of the dramatics department, in touring Europe this coming summer with Professor Pierre Macy's student group. It is a token of appreciation on their parts of the long term of diligent service she has given in that department.

The play is to be one of the regular annual commencement events.

Part of the cast was chosen by Miss Crowder in preliminary tryouts Thursday night. In taking the lead role as Eva, Lois Barall will make her first appearance on a Tulsa stage since her return from New York. Two years ago she was one of the outstanding students of dramatics at the University of Tulsa; and before that she took major roles in Central high school productions, as well as winning contests in dramatic reading.

Opposite her as Adam Smith, will be Grayden Coss, president of the graduating class, who carried the title role in "Clarence," played the husband in "Jane Clegg," a Tulsa Little Theater production, and has taken major roles in other plays both this year and last.

Other old-timers in the cast are Dolores Stikey and Leslie Seal. Dolores is a senior who has taken stellar roles since her days at Central high school. She carried the lead in "Daisy" this year, in "Thank You" last year, has written and produced a one-act play, "The Reporter," on the U. of T. stage, and has been in many other plays.

Leslie had a lead role in "Left That Off," and took part in "Daisy," "The Second Mrs. Tanqueray," and "The Mayor and the Manicure."

Wesley Henke, Matilda Frans and Mrs. Robert Wilson have also been given roles.

"Adam and Eva" was written by Guy Bolton and George Middleton. It was produced with great success in New York two years ago and has been successfully produced by Little theater groups throughout the country since then. It is a comedy in three acts.

Tulsa Dancing Team Are Making Good in Entertainment Field; Dance at Skirvin

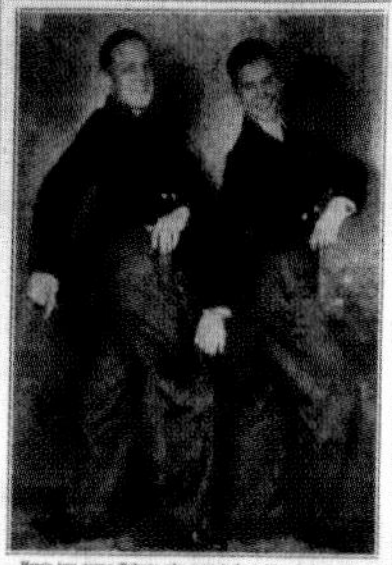

Here's two young Tulsans who seem to be getting Lady Luck's most golden smile. They're a young dance team, Ongring and Dana. Both are natives of Tulsa and were trained in Tulsa also, at the Gardiner Studios of The Dance. This week they have been dancing at the Skirvin hotel in Oklahoma City. Next week they have an engagement at the Bagdad Supper club in Dallas. The Grace Causey Enterprises is managing the boys.

Young Peoples and Local Actors Give One-Act Play Trio

The Young People's Society of Trinity Episcopal church and a group of young actors of the city combine next Tuesday evening to present three one-act plays. These will be given in the basement of the church and are "The Valiant," "Post-Mortems" and "Sham."

To the man who fears everything is probable.—Moose Magazine

Supper Club Transformed by Remodel Plan

Pratt's Orchestra to Be Final Touch

Skyland Gardens, the sixteenth floor of the Mayo Hotel is going through a process of re-decoration that when finished will make it one of the most beautifully arranged supper clubs in the west.

The old decorations and the old colonial fixtures are being removed and in their place are gorgeous beds and streamers of flowers. The windows will resemble the arches in the sunken garden.

Work will be completed and the opening has been announced for May 31, Mal Pratt and his orchestra returning to the hotel. They will entertain for luncheon, dinner and supper every day and night.

EASY CURE

The disease known as "inferiority complex" can readily be cured if the patient mingles freely with successful men—so called.—Moose Magazine

LIGHT COAT

A jaunty little outfit for day-time wear has a navy blue wool-flower skirt and a little semi-fitted and flaring jacket in beige featherweight wool.

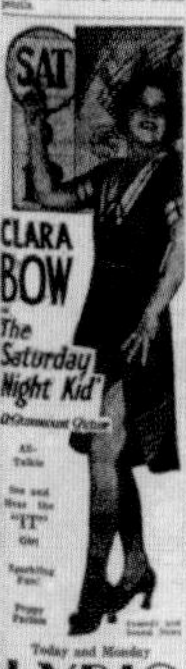

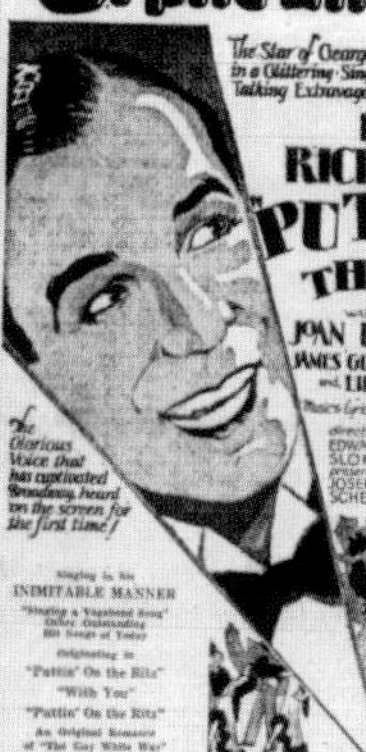

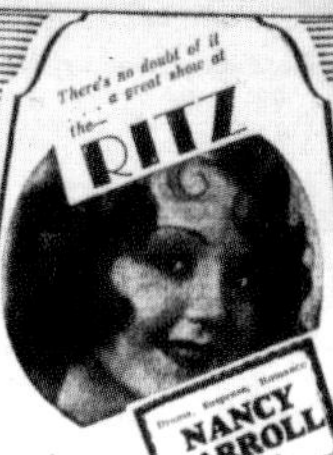

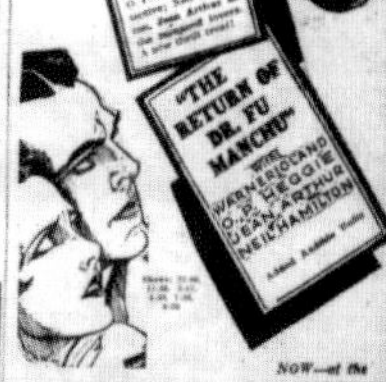

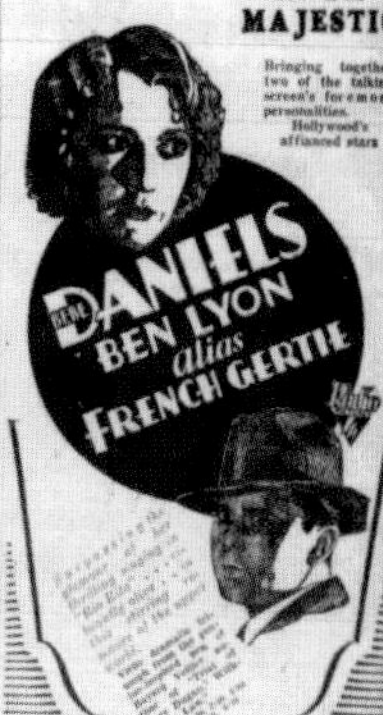

SELECTED PRESS
1930–61

"If you have a little spectrum in your home, playing the silent harmonic rhythms of abstract color and movement, all connubial disharmonies should be swept away, in the belief of Thomas Wilfred, who has adapted his famous 'color organ' to home use.

"His newest improvisation on the 'color music' idea is an instrument ranging in size from an average small radio, which could easily fit in a Greenwich Village hall bedroom, to a parlor sized machine. It plays 'color records' as a phonograph plays sound records and the wraith like driftings of the rainbow harmonies are observable upon a screen which resembles the sounding board of a loudspeaker space of a radio.

"One now sits and looks at symphonies and concertos, rather than listening to them.

"'It has the dream-like, reverie, quality of the old fireplace into which driftwood had been tossed, except that designs and restful pictures are created,' says the inventor."

—Gilbert Swan, *The Tulsa Tribune*,
May 18, 1930

The Tulsa Tribune, *Oklahoma, May 18, 1930, "Color Organ to Solve Problems of Ibsen Play: Invention Has Uses in Other Fields," Gilbert Swan*

New Musical Movement Launched by Artists

"Ether" Instrument Designers Sponsor New Radical Idea

NEW YORK, Dec. 20. (AP)—Hans Barth, composer and quartertone piano forte recitalist launched a project tonight to form an American society of exponents of unusual forms of music.

He presided at a conference with Maurice Martenot, inventor of so-called "ether" musical instruments. Leon Theremin, Russian inventor of the thereminvox and Thomas Wilfred Dane, originator of the Chavilux which projects color instead of sound on a screen when played.

An orchestra or ensemble will first be attempted.

Allentown Morning Call, *December 4, 1930, "New Musical Movement Launched by Artists"*

OPPOSITE: Iowa Press Citizen, *October 6, 1948, "Wallpaper May Cause Nervous Breakdown, Says Thomas Wilfred."*

Wallpaper May Cause Nervous Breakdown, Says Thomas Wilfred

Do you growl at children, bite your fingernails or braid your necktie? If so, the trouble may lie in the four walls of your living-room. Your wallpaper may be slowly and surely driving you mad without your knowing it.

The idea of wallpaper causing some nervous breakdowns is proposed by Thomas Wilfred, director of the Art Institute of Light, New York, who will demonstrate "light painting" at the University theater, October 19 at 8 p. m.

According to Wilfred, colors can be obnoxious to the extent of wreaking havoc with the human nervous system.

He claims that, although you may never be conscious of the color of your walls, the retina of your eye transmits the offending color to your nervous system, and before you know it you're flat on your back on a psychiatrist's couch.

Called the latest development in the field of art, Wilfred's "light painting" is done on a flat, white screen with a new type projection instrument.

Mobile visual compositions of abstract form and color are the results of Wilfred's experiments.

Tickets to the lecture-recital may be obtained at the ticket office in room 8-A, Schaeffer hall, or at memorial union. There will be no charge for the recital.

SAN FRANCISCO SUNDAY CHRONICLE January 4, 1959

Art Shows in the Eastern Galleries

By Alfred Frankenstein

BACK in New York, I spent an extraordinary morning with the only man alive who is not a painter, sculptor, or graphic artist but has, or at the time did have, a major work on display in each of the city's principal art museums — the Metropolitan, the Whitney, and the Modern. This was Thomas Wilfred, the inventor of Lumia, the art of mobile projected light.

Wilfred will be remembered for the recitals he gave here under the auspices of the San Francisco Museum on the keyboard instrument, called the clavilux, with which his name is especially associated. The clavilux is dismantled now, and Wilfred is devoting himself to the creation of what he calls recorded compositions. Both the Metropolitan and the Museum of Modern Art have added works of this kind to their permanent collections, and the Whitney had the latest one on loan.

These compositions employ color elements of stained glass, polished aluminum reflectors specially shaped to provide the forms, and simple but subtly balanced machinery to keep the whole in motion. They appear on a screen not much bigger than that of a large television set, but flat and standing vertically. The color cycles and the form cycles move at slightly different tempi. In the Whitney composition, entitled "Multidimensional," the color cycle takes six minutes and the form cycle seven, with the result that the work will run some 44 hours before it repeats itself. The Metropolitan's Wilfred, "Counterpoint in Space," also has a 44-hour time span, but the one at the Modern, "Vertical Sequence," runs 51 hours before it repeats. (It has been running, incidentally, for 18 years.) But every moment in these incredibly extended works is composed — that is to say, conceived in advance — and nothing is left to the accidental coincidence of moving lights and reflectors.

"Multidimensional" is the most complex of the three works I saw, but all are similar in their slow, majestic movement, their infinity of space, their sense of eternal flux, dissolving, and perpetual renewal, and their utterly unearthly calm. Many of the problems of modern painting reach a solution in Wilfred's work. It belongs with abstract expressionism in its most lyrical phases, and it relates to the new phase of abstraction wherein nature is reimplied if not reasserted. But what counts most of all is the serenity of spirit which these compositions create. I hereby nominate Thomas Wilfred for the Nobel Peace Prize.

San Francisco Sunday Chronicle, *January 4, 1959, "Art Shows in the Eastern Galleries," Alfred Frankenstein*

Inventor Of The Clavilux Plays Lumia In Church

Composer Of Music In Color Offers Recital In Park Ridge

By EVELYN PIZZA
(Staff Writer)

Park Ridge — Music was seen but not heard at the meeting of the Men's Fellowship of the First Congregational Church last night.

MEN'S CLUB GUEST

Guest speaker Thomas Wilfred, who conceived an art form named the lumia, presented his instrument for its expression, the clavilux.

The art form blends colors, motion, and light; it has been likened to music without sound. It has rhythm through the tempo of movement and depth through intensity of colors, yet there is no definite form nor pattern. Rather, there is a cloudlike flow of all possible shades of color.

Inventor Wilfred was born in Denmark in 1889, and has a background of music and art. He was a folksinger who played the 12-string archlute before he came to this country.

Here he continued experiments with light as an independent art medium, and in 1922 he offered his first lumia recital using an instrument which he called the clavilux. Since then he has given recitals here and abroad.

He said that although he had received acclaim, including a decoration in 1946 from Christian X of Denmark, he had suffered the trials of all pioneers.

However, he said, to be successful a pioneer must have confidence in his endeavor and in himself. To look at the nearly bald man who speaks meticulously, with just the slightest trace of an accent, it is obvious that he has that confidence and determination.

The instrument is in a console that looks like an oversized television, and it is automatic. A simple switch turns on the lumia composition.

In the base of the console, Wilfred explained, is a lamp directing a beam of light through transparent slides of colors reflected from lenses varied in shape and size to attain a desired effect.

The console is sealed, designed to continue indefinitely to produce the same visual themes and phrases. The Art Institute of Light in New York, a nonprofit research center for the development of lumia, makes the claviluxes. Wilfred is the head of the organization which he founded in 1930.

He is the sole creator of lumia compositions, and frequently composes his pieces on walks through the woods, he said. Each composition is an entity, not an interpretation of music or poetry.

When one watches the instrument for a few minutes, he may feel the effect of drifting in space. Serenity is evoked by the use of soft colors.

Yet each person's response will vary, he said, as it does in all art forms.

The clavilux has been used in psychiatric therapy to a small degree, he said.

The lumia viewed last night was "Multidimensional". The complete playing time is 44 hours, but it can be resumed at any point. It took about 3 months to make, and is priced at $3,400.

His largest piece of lumia was installed in Chicago in the Hotel Sherman in 1929. It is 21 by 210 feet, and is projected in a specially designed room. Wilfred said that after 22 years of constant use it still retains its original beauty.

He also has a clavilux on display in the Museum of Modern Art. In addition, Wilfred designs lighting equipment which projects scenery that can be used in place of painted flats.

Wilfred's talk was one of a series sponsored by the Men's Fellowship, and the last for this season. The Club will meet again in October.

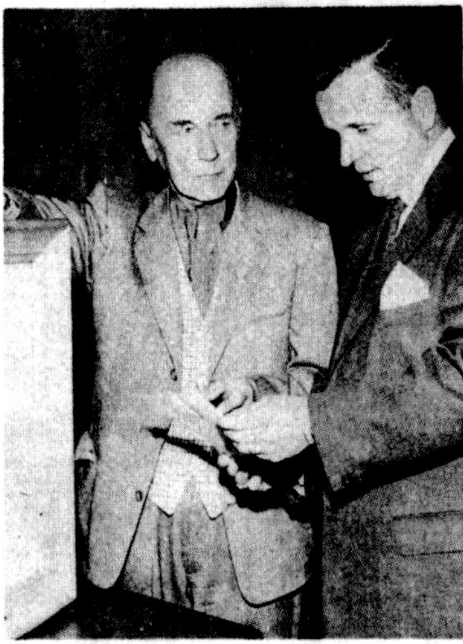

COMPOSER — INNOVATOR: Thomas Wilfred, left, stands beside his instrument, the clavilux, which looks like an oversized television console. Woolsey Serven, president of the Men's Fellowship of the First Congregational Church, reads the history of the art form the clavilux is designed to project. Wilfred was guest speaker at the Fellowship meeting. (The Record photograph.)

The Record, *May 4, 1961, "Inventor of the Clavilux Plays Lumia in Church," Evelyn Pizza*

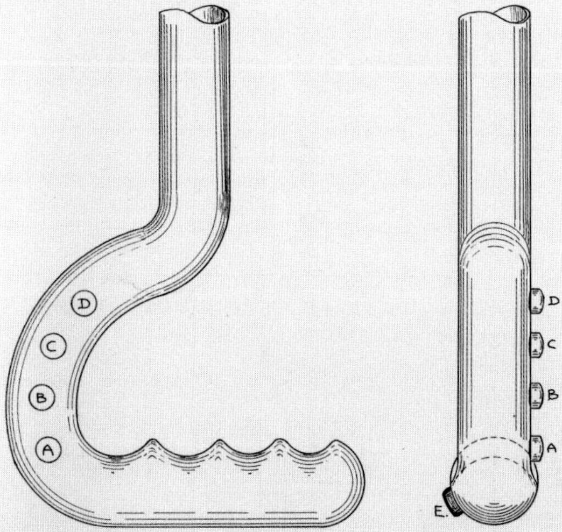

JOYSTICK HANDLE

137

FURTHER READING

Betancourt, Michael, ed. *Thomas Wilfred's Clavilux*. Borgo Press, 2006.

Brougher, Kerry, and Olivia Mattis, eds. *Visual Music: Synaesthesia in Art and Music since 1900*. Exh. cat. Thames and Hudson, 2005.

Duplaix, Sophie, and Marcella Lista, eds. *Sons et lumières: Une historie du son dansl'art du XXe siècle*. Exhibition catalogue. Centre Pompidou, 2004.

Orgeman, Keely, ed. *Lumia: Thomas Wilfred and the Art of Light*. Yale University Press, 2017.

Moritz, William. "Abstract Film and Color Music." In *The Spiritual in Art: Abstract Painting, 1890–1985*, ed. Maurice Tuchman et al., 297–311. Exhibition catalogue. Los Angeles County Museum of Art, 1986.

Stein, Donna M. *Thomas Wilfred: Lumia. A Retrospective Exhibition*. Exhibition catalogue. The Corcoran Gallery of Art, 1971.

Joystick Handle, *ca. 1940–1950*